CHROMATIN AND GENE REGULATIO
Mechanisms in Epigenetics

Chromatin and Gene Regulation
Mechanisms in Epigenetics

Bryan M. Turner
The Medical School
University of Birmingham
Birmingham, UK

Blackwell
Science

© 2001 by
Blackwell Science Ltd
Editorial Offices:
Osney Mead, Oxford OX2 0EL
25 John Street, London WC1N 2BS
23 Ainslie Place, Edinburgh EH3 6AJ
350 Main Street, Malden
  MA 02148-5018, USA
54 University Street, Carlton
  Victoria 3053, Australia
10, rue Casimir Delavigne
  75006 Paris, France

Other Editorial Offices:
Blackwell Wissenschafts-Verlag
  GmbH
Kurfürstendamm 57
10707 Berlin, Germany

Blackwell Science KK
MG Kodenmacho Building
7–10 Kodenmacho Nihombashi
Chuo-ku, Tokyo 104, Japan

Iowa State University Press
A Blackwell Science Company
2121 S. State Avenue
Ames, Iowa 50014-8300, USA

First published 2001

Set by Best-set Typesetter Ltd.,
Hong Kong
Printed and bound in Great Britain
by MPG Books, Ltd,
Bodmin, Cornwall.

A catalogue record for this title
is available from the British Library

ISBN 0-865-42743-7

Library of Congress
Cataloging-in-Publication Data

Turner, Bryan M.
    Chromatin and gene regulation :
    mechanisms in epigenetics /
    Bryan M. Turner.
      p. ; cm.
    Includes bibliographical
  references.
    ISBN 0-86542-743-7 (pbk.)
    1. Chromatin.   2. Genetic
  regulation.   I. Title.
    [DNLM: 1. Chromatin—
  genetics.   2. Gene Expression
  Regulation—physiology. QH 599
  T944c 2001] QH599 .T87 2001
  572.8'65—dc21
                    2001035095

DISTRIBUTORS

  Marston Book Services Ltd
  PO Box 269
  Abingdon, Oxon OX14 4YN
  (*Orders*: Tel: 01235 465500
        Fax: 01235 465555)

The Americas
  Blackwell Publishing
  c/o AIDC
  PO Box 20
  50 Winter Sport Lane
  Williston, VT 05495-0020
  (*Orders*: Tel: 800 216 2522
        Fax: 802 864 7626)

Australia
  Blackwell Science Pty Ltd
  54 University Street
  Carlton, Victoria 3053
  (*Orders*: Tel: 3 9347 0300
        Fax: 3 9347 5001)

The Blackwell Science logo is a
trade mark of Blackwell Science Ltd,
registered at the United Kingdom
Trade Marks Registry

For further information on
Blackwell Science, visit our website:
www.blackwell-science.com

For Matthew, Dominic and Ginny,
without whom nothing would make any sense

Contents

Preface

Early in the writing of this book, I decided to try and present the subject as a story. The story would have a beginning, a middle and an end (in that order) and would be accessible to the student and non-specialist. Unfortunately for this laudable objective, as the book was in progress, research into the ways in which DNA packaging, by proteins, regulates gene expression was growing at a truly terrifying rate, as it continues to do. The subject has seen something of an information explosion. Practitioners in many different areas of science and medicine have discovered that the DNA–protein complex packed into the nucleus of the cell (i.e. chromatin), is central to the problems they are studying. It is fundamentally important to cell growth control, cancer, immune system function and cloning, among many other fields, and is beginning to offer exciting new opportunities for therapeutic intervention. New results have not only added to our overall understanding, but have sometimes radically changed our ideas about the way things work. Constructing a story from such dynamic subject matter presented something of a challenge.

I have opted to start as simply as possible, with an overview of transcription in bacteria, which don't even use chromatin, but which do provide useful insights into how the problems faced by more complex organisms might be tackled. In succeeding chapters, I introduce a succession of new, and increasingly refined, structures and chromatin-based control mechanisms. This culminates, in the final chapter, in a discussion of dosage compensation, something that requires co-ordinated regulation of gene expression across an entire chromosome. It's not an end as such, and provides no final denouement, but it does bring together concepts and ideas that have been introduced in the earlier chapters. The chapters have been written as if they will be read in sequence, but this is not essential. Readers with a bit of background knowledge should be able to dip into selected topics without too much difficulty.

The basic structures by which DNA is packaged into the cell nucleus have changed very little over hundreds of millions of years of evolution. They are essentially the same in organisms from yeast to fruit flies to mice and men. All of these organisms offer their own advantages and disadvantages for experimentation, and they all appear at regular intervals throughout the book. Some of the matters we discuss may seem arcane at first, but those that are included are ones that have had a major impact on what we are usually most interested in, namely how our own bodies operate. On a practical note, the inclusion of results from such a wide range of organisms has created problems with terminology, particularly how the names of genes and their protein products are presented. Readers of this book need remember only one rule, (and this has been strictly adhered to), namely genes are in italics (e.g. *Aprt, SIR2*) and proteins are not (e.g. Aprt, Sir2). Capital and lower-case letters are often used in the scientific literature to distinguish genes and proteins, but conventions differ from one organism to another and, for our purposes, can be ignored.

Simon Rallison at Blackwell Science commissioned the book and encouraged me to start, and Anne Stanford nursed me through the early stages and gently cajoled as deadlines came and went. Various chapters have been read, sometimes in several different versions, by my friends Jim Allan (Edinburgh), Peter Becker (Munich), Steve Busby (Birmingham), Mitzi Kuroda (Houston) and Renato Paro (Heidelberg). Their insights, criticisms (always constructive, even when severe) and unfailing encouragement have kept me going and helped avoid embarrassing mistakes. I am very grateful. Needless to say, blunders that remain are entirely my own. While writing this book, I have tried to carry on working as an experimental scientist. Students, technicians and post-docs in the Chromatin and Gene Expression Group have tolerated my occasional preoccupation and memory lapses with good humour and forbearance. My colleague Laura O'Neill has taken on responsibilities beyond her years and never complained. Only through their collective efforts has our research enterprise stayed afloat.

Earlier this year, the chromatin community was shocked by the tragic death, in a road accident, of Alan Wolffe. Articles written by Alan, either reviews or research papers, are cited throughout this book, as is his own book on chromatin. His intellect, enthusiasm, willingness to help and sheer energy were inspirational. He was a good friend and will be greatly missed.

Bryan Turner
Birmingham
September 2001

Prologue

Comparisons with everyday things are often used in attempts to explain complex scientific problems. A well-chosen metaphor can produce that sudden flash of understanding that makes a teacher feel that their day has not been completely wasted. Biological scientists are particularly prone to metaphors. This is not because the problems they study are any more complex than those of other scientific disciplines (although they arguably are), or because the level of background knowledge among nonspecialists is any lower than in other areas of science. It is rather that the problems studied often impinge directly on our everyday lives in a rather personal way, either through their possible medical applications or because they deal with how we perceive ourselves and our place in the living world. There is a general desire to know and understand the intricacies of our biology, bringing with it an increasing need for specialists to explain themselves. In this respect, the science of genetics has taken centre stage. Genetics sets out to define the ways in which our lives are influenced by the biological legacy we inherit from our parents and to deduce our evolutionary origins. Such questions strike at the heart of what makes us human and unique.

Genetics has been revolutionized by the advent of new technologies with which the most complex of biological processes can be dissected at the molecular level. We now have techniques to manipulate genes with a precision and selectivity undreamed of a generation ago and which provide unprecedented experimental opportunities. The same techniques have catapulted genetics (not for the first time) into the public domain. Genetically engineered plants and animals are now in commercial use, gene therapy is beginning to establish its clinical potential and the possibility of selective genetic manipulation of the early human embryo is an obvious, though not necessarily desirable, extension of now routine *in vitro* fertilization technology. Genetics rests firmly in the public eye.

In view of this, it is hardly surprising that books about genes and genetics feature prominently in the best-seller lists. Some are good and a few are very good indeed, successfully bridging the gap between the specialist geneticist and the curious and concerned nonspecialist. But what they all tend to do, sometimes unconsciously and sometimes explicitly, is to emphasize the *power* of genes. Genes, it is sometimes claimed, make us what we are. They determine our appearance, our health, our intelligence and our behaviour. We only need to identify the appropriate gene or genes, adjust as necessary with our new technologies, and anything is possible. This scenario is exciting and frightening, but is it justified? Perhaps before getting too carried away by the undoubted

1

importance of our genes, we should step back and ask exactly what they can and cannot do.

The helpless gene

Of the various macromolecules that make up our cells it is, with rare though important exceptions, the proteins that actually do things. Actin, myosin and their associated muscle proteins harness chemical energy to provide the motive power for cells and muscles, proteins embedded in cell membranes allow our cells to receive and interpret signals from the environment, while myriad enzymes catalyse the chemical reactions that make up essential metabolic pathways. Proteins will work even in isolation outside the cell. Given the right conditions of pH and temperature and the right substrates and cofactors, an enzyme will catalyse a chemical change in a test tube, or even in a washing machine as a component of a 'biological' soap powder. In contrast, a gene in a test tube will do nothing at all. It has no catalytic ability. The average gene is just a rather long, but chemically simple, macromolecule, which is no more complex, and certainly less useful on its own, than the protein that makes up the wool in a sweater.

This may seem at first sight a rather trivial point and unfair besides. It's surely not necessary for genes to do things themselves. Their power and influence is exerted indirectly through the information they contain. To address this point, we can borrow a particularly appealing metaphor from Richard Dawkins (*The Blind Watchmaker*, Longman 1986) that compares our genes to the recipe used to bake a cake. Like the recipe, genes are simply a set of instructions that must be followed to achieve a specified end. So what should we expect of a recipe? Well, clearly we can't expect it to do much on its own. Even if we leave our cookery book open at the right page in a well-equipped kitchen with all the right ingredients close by, a cake is unlikely to appear. Unless, that is, a sympathetic friend who can take a hint happens to enter the kitchen. Then a cake may materialize. So, to follow the metaphor dangerously far, we must now ask what the sympathetic friend has added to the situation that has made the cake possible, apart from a kind heart? The answer is information *beyond what is contained in the recipe itself*. Information, for example, that enables him or her to read and interpret the recipe in the first place and the learned skills to carry out the tasks that it specifies.

Let's leave the kitchen for a bit and ask what conditions *genes* require in order that the information they contain can be converted into a living organism? The genes will require the molecular apparatus for transcribing DNA into RNA and translating RNA into proteins, together with *all the structures and metabolic processes* that enable these events to proceed in an orderly and coordinated fashion such that a living cell can be assembled. All this is found only in a pre-existing cell. But even this is not enough. In the animal kingdom, in order to progress from a complete set of genes to a living organism, not just any cell will do. Only the fertilized egg can enable this progression to occur, the reason being that only this cell contains the information that specifies the sequence of genetic events driving early development. This information usually takes the form of proteins that activate or suppress particular genes or gene families and that are often distributed in a specific way throughout the cell. This distribution generates protein concentration gradients that not only impart positional information, but also lead to different patterns of gene expression in different parts of the developing embryo. All this information is an integral part of the structure of the egg. It is nothing to do with the egg's DNA. In fact, the idea that factors outside the genome could be of overriding importance in determining gene expression is a rather old one. The classic experiments by Henry Harris and colleagues in the 1950s showed that, in hybrid cells, the cytoplasm rather than the nucleus was the more important determinant of patterns of gene expression.

The primary purpose of this introduction is to emphasize two simple but crucial concepts. The first is that most genes do what they are told. Genes have only the *potential* to be transcribed (switched on). In many cases, whether a gene is on or off will depend on the cell in which it finds itself and on the cells in which it has been, in other words on its genetic history. It is true that some genes, the so-called *housekeeping* genes, are always transcribed. These genes encode enzymes or structural proteins and RNAs that are essential for all living cells, so the fact that they are always on is not surprising and certainly does not constitute evidence that genes are immune to cellular controls. It is also true that just a single gene, when mutated, can exert a major, sometimes catastrophic, effect on the individual. But, even mutated genes often exert an effect only when their information, or misinformation, is interpreted by the molecular apparatus of the cell and converted into an abnormal effector molecule, usually a protein. The point is worth bearing in mind because, as discussed later in this book, the steps that lead from mutated DNA to altered protein often provide a more useful target for therapeutic intervention than the gene itself.

The second point is that, in the animal kingdom, we inherit more from our parents, our mothers especially, than just a bundle of DNA. (The continuing restriction of certain comments to the animal kingdom is because many *plant* cells are 'totipotent', which means that, with the right growth factors, a complete plant can be grown from a single somatic cell.) The egg contains a store of information essential for the correct initiation of genetic activity in the early zygote and without which the DNA can no more produce a living organism than our recipe can produce a cake. In addition, both parents can influence the way that some genes are expressed in the developing embryo. A small proportion of genes, said to be 'imprinted', are switched on or off in the embryo depending on whether they were inherited from the father or the mother. The imprint does not reside in the DNA sequence, the same gene can be on in one generation and off in the next, but instead is a chemical or structural marker, the nature of which is as yet unknown, that is placed on the gene by the germ cell (sperm or ovum) in which it was last resident. Thus, the characteristics of any given cell are an amalgam of genetic (via DNA), epigenetic (inherited information apart from that encoded in DNA) and environmental influences. To ask which is more important in producing a functioning cell is as futile as asking whether flour or currants or a working oven are more important for a successful fruit cake.

Different cell types have the same genes, but differently expressed

The initial choices as to which genes to express are made during development.

We all start as a single cell, the fertilized egg. As development proceeds and cell numbers increase, the choices open to any particular cell diminish progressively, culminating, for many cells, in progression down a pathway of terminal differentiation. This term defines that stage of a developmental pathway when choice is gone and cells' fates are sealed. One of the most valuable insights to come out of research in the biological sciences over the past 40 years is that, throughout this progressive restriction of choice, each of our cells retains a complete set of genes. The genetic recipe unique to the individual organism is present, with only minor exceptions, in *every one of its cells*. This understanding grew initially from experiments carried out in the 1950s with developing frog embryos, which showed that a nucleus transplanted from a body cell of a developing embryo into a fertilized egg from which the nucleus had been removed could go on to develop into a perfectly normal adult frog. These experiments gave the first indication that the genetic changes occurring through development were not irreversible. This conclusion is valid even for mammals, as demonstrated by the cloning of sheep, cattle and mice by transplanting nuclei from mature cells into fertilized eggs. Cloning technology is still far from perfect, and may never become so, but even as they stand, the results provide a dramatic demonstration of the ability of the fertilized egg to tell the genes placed under its control exactly how to behave.

So, cells are what they are, not because of the genes they *have*, but because of the genes they *express*, i.e. which genes are on and which are off. In order to understand what makes cells different and how they change from one type to another during development, we must understand the mechanisms by which this is achieved. This problem, so simply stated, is the most fundamental and complex in modern molecular biology.

Defining the cell's gene regulation requirements

It is a useful exercise to try and define, in broad terms, the tasks that a cell must accomplish in order to regulate its genes appropriately. Such definitions are always imperfect, and can become a burden if we forget that they are artificial constructs designed only to guide our thoughts.

First, every cell must be able to switch certain of its genes on and off, usually in response to signals from outside the nucleus, as part of its day-to-day routine. For example, all cells switch specific genes on and off at particular stages of their growth and division cycle. Indeed, the presence or absence of the protein products of these genes is a central element of growth control. Second, all cells have genetic mechanisms by which they attempt to cope with unexpected environmental changes, such as sudden shifts in temperature or the appearance of toxic chemicals. A family of genes called the stress response genes is activated rapidly by a variety of potentially damaging environmental insults. Genetic switches are an integral part of the most fundamental elements of cellular existence, not only growth, replication and survival, but virtually any other one cares to consider.

Gene switching is also essential for cells to change from one type to another during development and thereby generate the various different cell types found in multicellular organisms. Some genes are transcribed only in certain cells and their protein products establish the very characteristics that define that cell type. The β-globin gene, expressed only in erythroid cells, is a good example. But such high-profile genes are not the only ones that fall into this category. Others, working behind the scenes, may be more influential, often operating to regulate the activity of other genes and gene families. Such genes may initiate whole developmental programmes rather than just encoding the products characteristic of the differentiated cell. They are themselves part of a switching mechanism.

It is a little appreciated fact that, in any particular cell type, most genes are permanently switched off. The process by which this is done has come to be referred to as silencing and it is of fundamental importance. To appreciate this, one has only to think of the numbers of genes involved. Somewhere in the region of 25 000 genes (i.e. 2/3 or more of all our genes) are *not* transcribed in any particular cell in the human body and must be kept quiet. Even a small amount of transcription (a whisper, say) from each of them would, in total, amount to sufficient background noise to drown out the cell's essential signalling mechanisms. Conversely, there are certain processes that all cells must keep running in order to survive. These include the various metabolic pathways required for energy generation and for the synthesis of essential metabolites, proteins and nucleic acids. The enzymes, structural proteins and nucleic acids required for these essential processes are encoded by a set of genes collectively known as housekeeping genes, a term that aptly describes their simultaneous lack of glamour and fundamental importance in the day-to-day life of the cell. These genes are always on.

Having established patterns of gene expression that define the body's different cell types, mechanisms must be put in place to maintain these patterns from one cell generation to the next. Cells must be able to remember who they are. This is a far from trivial problem. The cell's growth and division cycle involves dismantling the careful, higher-order packaging of its DNA, stripping away DNA-associated proteins, making an exact copy through the action of DNA polymerases and associated enzymes and then repacking the two copies into two separate daughter nuclei. What markers can survive this trauma so that the two daughter cells, unless instructed to do otherwise, both show just the same pattern of gene expression (i.e. are of the same cell type) as their parent?

An important point to take away from this definition of the overall problem is that there are different kinds of gene regulation, each of which fills a different functional need and each of which may be brought about by very different molecular mechanisms. To be sure, there may be common themes and common mechanisms – it would be surprising if there were not – but to speak of mechanisms of gene regulation in general terms, without specifying exactly what sort, is a dangerous simplification.

In the following chapters, I attempt to outline the ways in which eukaryotic cells (i.e. cells with nuclei) make decisions as to which genes to express and how the patterns of gene expression characteristic of given cell types are set and maintained. A recurring theme throughout the book is that genes cannot be considered in isolation. Their activities are part of an integrated programme of growth and development carried on within a defined subcellular structure (i.e. the cell nucleus). Like players in an orchestra, they must stick to the score. If the genes are musicians, the cell is a classical symphony rather than a piece of traditional jazz. Which brings us, finally, to the central theme of this book, namely the nature of the integrating mechanisms that both regulate and co-ordinate the genetic activities of our cells.

The integrating function of chromatin

DNA is packaged into the nuclei of eukaryotic cells by proteins, the most numerous of which are the histones. The DNA–protein complex within the nucleus is called chromatin. Details of this are presented in Chapter 3, but it is worthwhile emphasizing the complexity of the packaging problem with which eukaryotic cells are faced. Let us take ourselves as an example. Every nucleated human cell contains about 2 m of DNA. A typical cell nucleus into which this DNA is constrained is about $10\,\mu m$ (i.e. $10^{-6}\,m$) in diameter. To appreciate the magnitude of the problem it is useful to scale things up to a more easily visualized size. So let's imagine that the nucleus is 10 cm in diameter (i.e. 10 000 times its actual size), about as big as a large grapefruit. The DNA to be packaged into our grapefruit is now 20 000 m (i.e. 20 km) long. We must also bear in mind that the DNA confined in our grapefruit must be able to function. It must be replicated and its genes transcribed in the way appropriate to the cell in which it finds itself. Reasoning from first principles, it seems extremely unlikely that cells would have evolved systems to accomplish this extraordinary feat of packaging that were not, at the same time, used to regulate the function of DNA. Indeed, it is difficult to imagine a packaging system that would not, inevitably, exert a major functional effect. There is now a rapidly growing body of experimental evidence to show that packaging of DNA and its organization within the cell nucleus play central, sometimes dominant, roles in regulating its function. It is this relationship between DNA packaging and gene expression that the rest of this book addresses.

Chapter 1: Controlling Transcription: Shared Aims and Common Mechanisms

Introduction

Even the simplest free-living organism must be able to deal with the inevitable changes in environmental conditions that it will encounter. These include changes in temperature, pH, availability of nutrients and so on. Metabolic pathways that are appropriate for one set of conditions may be unnecessary, or even disadvantageous, in another. We now know that organisms usually meet these environmental demands through regulating the expression of specific genes. For example, if an organism is living in an environment that (temporarily) lacks a source of the sugar galactose, it makes little energetic sense to synthesize the enzymes required for galactose metabolism. It is more efficient to switch off the genes making these enzymes, and switch them on again when conditions change and galactose returns to the environment. Of course, like everything else, this option is not without cost. It requires that the organism has in place mechanisms that can switch specific genes on and off in response to environmental and metabolic demands. As we will see, these mechanisms require specific proteins and synthesis of these proteins itself places demands on the organism's energy resources. However, the alternative strategy, namely for all genes to be switched on all the time, has not proved evolutionarily popular. It is certainly simple, but is not (apparently) cost-effective.

Over the past several years, it has become clear that the fundamental mechanisms of gene expression and regulation that have evolved in order to meet fluctuating metabolic demands show a remarkable degree of conservation from one organism to another. This is something of a bonus for experimental scientists. We can study relatively simple model systems, reasonably secure in the knowledge that at least some of the mechanisms that are uncovered will not only be useful and interesting in themselves, but will also be relevant to more complex situations. In this spirit, I will start off by stating some general principles and then begin exploring the enzymology of transcription from the point of view of some of the more simple (although evolutionarily most successful) organisms, the bacteria.

Some general principles

The Central Dogma

Since the early 1960s, it has been accepted that genes and the proteins encoded by them are linked by an 'unstable intermediate' now called messenger RNA (mRNA). The simple, unidirectional pathway DNA → RNA → protein has come to be known as the Central Dogma. Like most dogmas it is not *completely* true. An enzyme encoded by retroviruses (viruses that use RNA rather than DNA in their genomes) called reverse transcriptase, can make DNA on an RNA template. But no example has yet been found of a system whereby a protein is used to code for RNA or DNA. The basics of *transcription* (synthesis of RNA on a DNA template) and *translation* (synthesis of proteins from an mRNA template) are shown diagrammatically in Fig. 1.1. Some basic information on the structure of DNA is presented in Box 1 at the end of this chapter.

Potential control points

The ultimate purpose of gene regulation is usually to change the amount of a particular protein or group of proteins. It is, after all, the proteins that perform most of the functions that the cell needs to stay alive. This being so, it is apparent from even a simple diagram such as Fig. 1.1 that there are several points along the pathway from gene to protein at which control might be exercised. These may be listed as follows:

1 *Initiation of transcription.* Control can be exercised by varying the frequency with which a polymerase molecule binds to the gene and begins transcribing. This is the first control point and, usually, the one that sets the agenda for all that may follow. If the gene is 'off' at this stage, then all other controls become redundant.

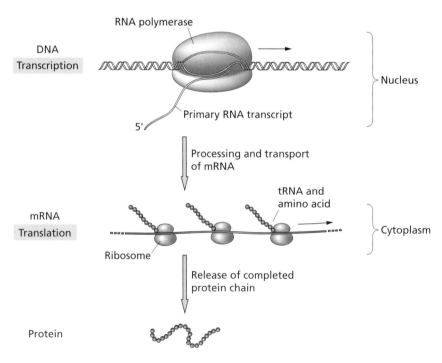

Fig. 1.1 The pathway from gene to protein.

2 *Rate of transcription*, i.e. the speed at which the polymerase progresses along the gene. At first sight this is not a likely control point. Enzyme reactions are dependent on variables such as temperature, substrate concentration, pH, etc., some of which are outside the cell's control and none of which are obvious candidates for regulating the transcription of specific genes. However, there are proteins that are known to have an essential role in the elongation stage of transcription. Several such proteins have recently been identified and characterized in eukaryotic cells. Some have a general role and others are required only for certain genes. Such proteins may turn out to have important roles in transcriptional control.

3 *Premature termination of transcription*. This is known to occur in bacteria and has been given the name 'attenuation'. It is intrinsically wasteful (the partial transcripts are destroyed) and seems to be used for fine-tuning patterns of expression, for example in eliminating small amounts of transcription from genes that have already been turned down.

4 *mRNA stability and processing*. mRNAs differ widely in their stabilities within the cell, often for good reason. A protein that is needed only briefly is best made by an unstable message so that its synthesis rapidly declines once the gene is switched off. (A good example of control through variable stability is provided by the RNA product of the mouse *Xist* gene; this is discussed in Chapter 11.) Stability is partly determined, in a still rather poorly defined way, by noncoding regions of the RNA molecule. In higher organisms, many RNA transcripts are modified as they travel from their sites of synthesis on DNA to the sites at which they are translated. In eukaryotic cells this can, in some cases, involve a process of cutting and pasting of the primary transcript (known as *splicing*) that can result in different cytoplasmic mRNAs and hence the synthesis of different proteins from the same gene. We return to this topic in Chapter 2.

5 *Translational controls*. mRNA may be sequestered or translation may be selectively blocked. For example, histone mRNAs are sequestered in the pronucleus of the sea urchin oocyte and released into the cytoplasm only after fertilization. There is also an increase in the translation of pre-existing oocyte mRNAs following fertilization, probably through the operation of several control pathways.

This brief and certainly incomplete list is intended only to emphasize the point that control of transcription *initiation* is just the first in a hierarchy of control mechanisms that cells use to regulate the amounts of specific proteins or nonmessenger RNAs. However, as we will see, controls on initiation are often of over-riding importance.

Transcription in prokaryotes

Bacterial RNA polymerase

RNA polymerase is the first component of the pathway leading from a gene to a functional protein. The enzyme must, in roughly this order, recognize the DNA sequence to which it must bind (more of this later), dissociate the two DNA strands (only one of them will act as the 'coding' strand), align ribonucleoside triphosphates with their complementary bases on the coding strand, catalyse the formation of the sugar–phosphate backbone of the growing RNA chain, and progress along the DNA. All of these abilities are essential for a successful polymerase and, indeed, for a viable cell. But as we are concerned with the regulation of transcription, and as regulation is achieved primarily by influencing the binding of polymerase to DNA, we will focus our attention on this first step.

The basic structure of the RNA polymerase from the small bacterium *Escherichia coli* (*E. coli*) is shown diagrammatically in Fig. 1.2. The major form of the enzyme has a molecular mass of 465 kDa and contains four different subunits designated α (of which there are two copies), β, β′ and σ. At first sight this might seem a rather complex structure, but considering the task the enzyme must accomplish, it is a model of simplicity.

Bacterial promoters

RNA polymerase must be able to recognize and bind to DNA sequences upstream of genes in such a way that it is precisely positioned to initiate transcription at exactly the right base. The DNA sequence adjacent to the transcription start site, to which the polymerase and other proteins can bind, is known as the *promoter*. Its recognition is the task of the σ subunit. (There are seven different σ subunits in *E. coli*, of which the most frequently used is σ70.) It is often the case in bacteria that a single promoter controls several structural (i.e. protein coding) genes. The polymerase passes through all genes, making a single, continuous RNA transcript. The task of making separate proteins from this transcript is delegated to the protein synthesis machinery. The linked structural genes usually encode proteins that work in the same metabolic pathway, so it makes sense for them to be regulated together. The whole coordinated unit, i.e. promoter, structural genes and any additional control regions, is known as an *operon*. The example shown in Fig. 1.3, the *lac* operon, contains three structural genes, all encoding enzymes involved in lactose metabolism, along with a promoter, a gene encoding a protein that can shut down transcription (a *repressor*) and a site for the repressor protein to bind to (the *operator*). We will deal with regulation of the *lac* operon in a later section.

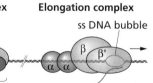

Fig. 1.2 Subunit composition of prokaryotic RNA polymerase.

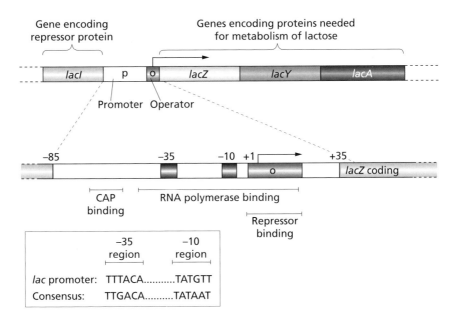

Fig. 1.3 The *lac* operon. The arrow shows the transcription start site and direction of transcription. Each RNA transcript incorporates the *lacZ*, *lacY* and *lacA* genes in a single length of RNA. The expanded version of the promoter region shows the extent of protein binding sites. DNA bases 'upstream' (5′) of the first base to be transcribed (assigned the number +1) are, by convention, given negative numbers and those 'downstream' (3′) are positive. Sequences characteristic of the two regions crucial for binding of the σ subunit of RNA polymerase (at around –35 and –10) are shown. The consensus sequence is made up of the bases found most often at each position. The sequences of individual promoters usually differ slightly from the consensus sequence.

Promotor sequences in *E. coli* and other bacteria are diverse. Some are loosely described as 'strong' (i.e. they bind the polymerase with *high affinity* and initiate *frequently in vivo*) and some are said to be 'weak' (i.e. they bind the polymerase with *low affinity* and initiate rarely). Despite these differences, promoters recognized by the σ70 subunit of RNA polymerase do bear some resemblance to one another. In particular, there are two six-base sequences positioned at around 10 and 35 base pairs upstream from the transcription start site (i.e. at positions –10 and –35) that vary relatively little from one promoter to another. These sequences are marked in the example promoter shown in Fig. 1.3. The *'consensus'* sequences for these two regions are TATAAT (–10) and TTGACA (–35), by which we mean that the great majority of bacterial promoters have sequences that are related to them. The promoter shown in Fig. 1.3 (the *lac* promoter) is typical in that it contains, in addition to the –10 and –35 sequences, an additional region for binding of a protein (CAP) that is involved in regulation of transcription (see below).

In mutant bacteria in which one of the bases within a conserved region has been replaced by another, binding of RNA polymerase will be altered. Substitution of bases within the nonconserved, intervening region has virtually no effect, but lengthening or shortening this region by insertion or deletion of bases seriously disrupts polymerase binding. The explanation for this lies in the way in which the polymerase binds to the promoter. Defined regions of the σ protein contact the DNA at both the −10 and −35 sites and these contacts have a certain degree of base specificity (i.e. only some combinations of bases allow binding to occur). The intervening DNA keeps the two conserved regions the right distance apart to allow the appropriate contacts to be made.

No single base is *absolutely* required in all σ70 promoters. This suggests that what the protein recognizes and can bind to is a combination of DNA sequence and conformation (shape), rather than just a specific array of bases (the way in which proteins recognize specific DNA base sequences is described later). It seems that several different combinations of bases can each provide sufficient DNA–protein bonds to allow binding. This rather permissive binding specificity is in sharp contrast to the very stringent specificities shown by other DNA-binding proteins. On the other hand, σ70 must bind to a variety of different promoters. If they all had the same sequence, then it would bind to them all with the same affinity and, other things being equal, initiate them all at the same rate. This is not what is needed at all. Some genes must be frequently transcribed, whereas others are transcribed infrequently or only under certain conditions. In the latter two cases, a *weak* promoter is what is needed. In fact, mutations that convert a weak promoter into a stronger one are often deleterious to the cell. Therefore, it is a mistake to think of weak promoters as second rate or ineffectual, simply waiting for evolution to improve their performance. They fulfil specific and crucial roles.

The role of σ factors in promoter recognition

Strong and weak promoters constitute the first level of gene regulation in bacteria, defining in the broadest terms the frequencies with which different genes will be transcribed. The role of σ factors is primarily in initiating transcription through the correct positioning of the polymerase within promoters. σ70 is the most common sigma factor in *E. coli*, but there are six others, each of which can substitute for σ70 in the RNA polymerase 'holoenzyme' (i.e. the functional polymerase complex with both catalytic and regulatory subunits). Each σ factor confers on the polymerase the ability to recognize a particular promoter consensus sequence and thereby initiate transcription of a defined family of genes.

There is evidence that σ factors also play a role in the separation of DNA strands ('melting') at around the −10 position. Strand separation, also referred to as the transition from the 'closed' to the 'open' complex, is a necessary precursor of transcription. However, the σ factor dissociates from the transcription complex after about 10 bases have been transcribed, so it is clearly not needed for continuing DNA strand separation or for any of the catalytic steps of RNA synthesis or polymerase progression. It can be regarded as a transcription initiation factor.

In some cases, the rate of transcription achievable by σ70 binding needs to be increased still further. The genes encoding RNAs that form part of the ribosomal subunits, the rRNA genes, are the most rapidly transcribed in *E. coli* and have the strongest promoters. But, in addition, they have sequences 40–60 bases upstream of the transcription start site that bind to the α subunits of the RNA polymerase holoenzyme, which are necessary for their very high rates of transcription. Mutations in the α subunits reduce transcription of rRNA genes but have little effect on transcription of other genes. Of the five subunits in the RNA polymerase holoenzyme, three (one σ and two α) are concerned primarily with regulating initiation through binding to selected DNA sequences.

Although control through differing promoter strengths and a choice of σ factors is a useful first step, it cannot meet all the needs of even a relatively simple cell. A successful bacterium must be able to adapt continually to changes in its environment, particularly the changing availability of specific nutrients. It does this by making the appropriate enzymes, or shutting down manufacture of those it doesn't need. This continual adjustment in response to environmental cues requires additional controls.

Genetic switches in bacteria

Ligand-dependent DNA-binding proteins

The switching mechanisms discussed here are made possible by proteins that have two binding abilities. First, they can bind DNA in a sequence-specific manner, and second, they bind a potentially useful metabolite, an amino acid such as tryptophan or a sugar such as lactose. In this context, the bound molecule is referred to as the ligand. What makes these proteins special is that, having bound their ligand, they undergo a conformational change that influences, positively or negatively, their ability to bind DNA. This is illustrated in Fig. 1.4. Proteins that bind to DNA sometimes inhibit transcription, most simply by blocking the promoter region to which the polymerase binds. Alternatively, a DNA-binding protein can facilitate the binding of polymerase to a weak promoter, perhaps by binding directly to one of its subunits and holding it in place. Thus, transcription of selected genes can be either increased or decreased in the presence or absence of specific ligands. The *lac* operon in *E. coli* provides a good example of how this works in practice.

Ligand-dependent regulation of the *lac* operon

The *lac* operon (Fig. 1.3) is regulated by both an activator and a repressor, each of which functions in a ligand-dependent fashion. When lactose levels are low, the *lac* repressor (with no bound lactose) binds to an operator site next to the *lac*

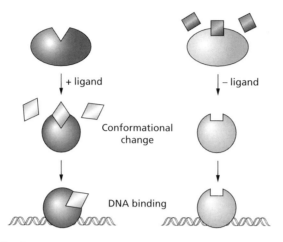

Fig. 1.4 Ligand binding can change the conformation of some proteins and their ability to bind to DNA. Some proteins can bind small molecules (e.g. sugars such as galactose or steroids such as retinoic acid) at specific sites. Binding of the 'ligand' (◇ and ■ in the diagram) can cause the protein to change shape (shown as round or oval in the diagram). Some proteins have binding sites for both small molecules and DNA. Some of these proteins adopt a shape that enables them to bind DNA only when they are also bound to a specific ligand (left-hand side). Others can adopt such a shape only when *not* bound to a ligand (right-hand side).

promoter and prevents transcription of the genes encoding enzymes necessary for lactose metabolism. When lactose levels increase, an increasing number of *lac* repressor proteins bind the sugar. The affinity of the ligand-bound repressor for operator DNA is about 1000-fold lower than that of the unliganded (i.e. lactose-free) repressor, with the result that increasing numbers of operators become free of repressor and transcription can begin (Fig. 1.5). While prevention of repressor binding is *necessary* for transcription to begin, it is not *sufficient*. Initiation of transcription requires, in addition, that an *activator* protein binds to the promoter just upstream of the RNA polymerase binding site. The binding of this activator is also ligand-dependent. The ligand in this case is the nucleotide cyclic AMP (cAMP). The possible protein–promoter combinations are shown in Fig. 1.5(a–d).

As usual, there is a good biological reason for this control mechanism. β-galactosidase converts lactose to glucose and galactose (which is then itself converted to glucose). Glucose is the preferred (most efficient) energy source for bacteria, so when glucose is present in the growth medium, it makes sense for the genes for metabolizing lactose to remain switched *off*, whether or not lactose is present in the medium. This has been achieved by making transcription of these genes dependent on an activator protein that binds to its operator site only when glucose levels are *low*. This could have been accomplished by making activator binding sensitive to glucose itself, but in fact it is regulated by a chemical whose level *increases* as glucose levels *fall*, namely cAMP. This reagent is also an important second messenger in eukaryotic cells and a major player in the

Proteins bound to promoter region	Lactose	Glucose	cAMP	*lacZ* transcription
(a) 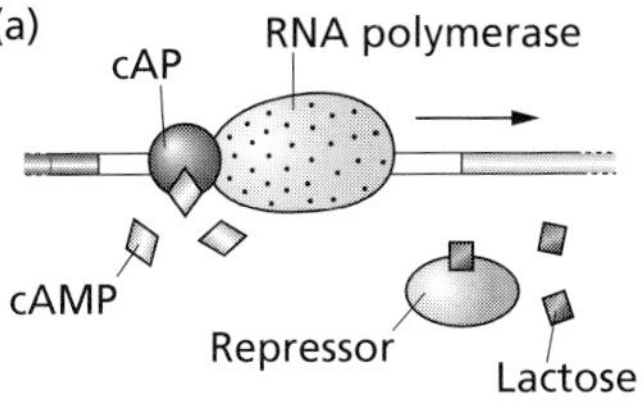 cAP RNA polymerase cAMP Repressor Lactose	+	−	+	Yes
(b) 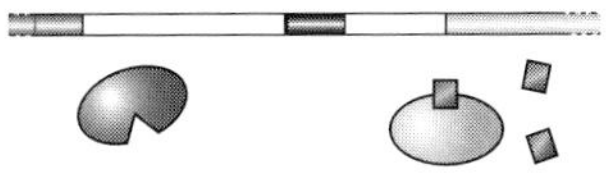	+	+	−	No
(c) 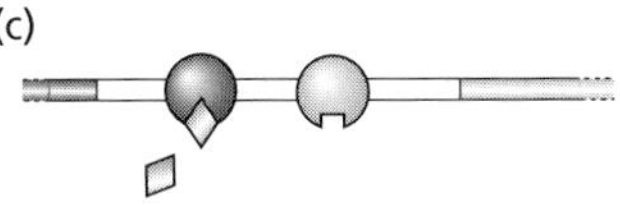	−	−	+	No
(d) 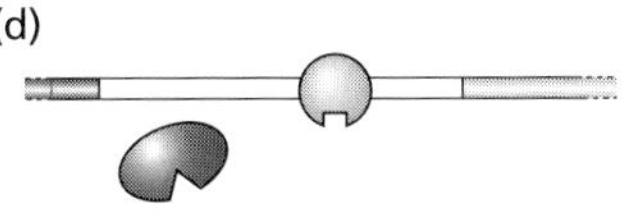	−	+	−	No

Fig. 1.5 Regulation of *lacZ* transcription by lactose and cyclic AMP.

regulation of some eukaryotic genes. The activator protein is known as CAP (catabolite activator protein, a name that pre-dates knowledge of the role of cAMP in its function). At first sight, the use of cAMP rather than glucose itself in this circuit seems an unnecessary complication. It is probably done because a fall in glucose needs to allow the activation not just of the genes necessary for lactose metabolism, but also of those needed for use of other carbon sources. The circuit must be able to integrate a complex set of metabolic demands and cAMP rather than glucose may be better suited to this. Whatever the explanation, the fact is that if glucose levels are low and cAMP levels are high, then the activator–cAMP complex will bind to a site adjacent to the *lac* promoter and facilitate RNA polymerase binding (Fig. 1.5a). It has been shown that this occurs by direct contact between CAP–cAMP and the polymerase.

DNA–protein interactions

Transcription and its regulation are governed by DNA–protein interactions of one sort or another. As we have already seen, DNA-binding proteins determine whether a gene is transcribed or not, and, if it is, at what rate transcription proceeds. Over the past few years, various techniques of structural analysis have provided a wealth of new information on the structure of DNA-binding proteins. Fortunately, some valuable simplifying principles have emerged.

DNA-binding proteins, irrespective of their origins, usually fall into one or another of four families, each of which is identified by a characteristic DNA-binding structural motif. The distinguishing motifs are shown in Fig. 1.6. Members of each of these families crop up in different parts of the book. In each case, DNA binding is brought about by the juxtaposition of specific amino acid side-chains with the edges of base pairs within the DNA helix. This is usually achieved by inserting part of an α-helical region of the protein into the major groove of the DNA. This initial binding step can then cause structural changes in both DNA and protein, leading to additional interactions that stabilize the binding. Most of the DNA-binding proteins encountered so far have been members of the helix-turn-helix (HTH) family and will be used to

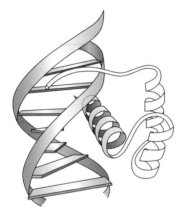

(a) Homeodomain

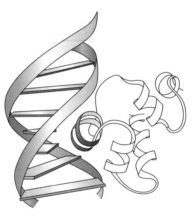

(b) λ-repressor (helix-turn-helix)

(c) Zn-finger

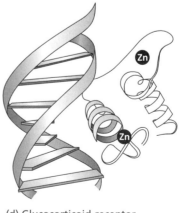

(d) Glucocorticoid-receptor

Fig. 1.6 Structural motifs that characterize four families of DNA-binding proteins.

illustrate some general points about the nature of DNA–protein binding and how it can be regulated.

Recognition of DNA by helix-turn-helix proteins

The DNA-binding region of HTH proteins consists of two α-helical regions separated by a short, unstructured region (the turn). The whole motif usually comprises about 20 amino acids, although the way these are partitioned between the helices and the turn can vary. The two helices of a typical HTH protein have rather different functions. One, the more carboxyl-terminal, has been shown by structural analysis of a number of DNA–HTH protein complexes to lie within the major groove of DNA, where some of its amino acid side-chains can form hydrogen bonds (H-bonds) with the edges of DNA base pairs. The λ-repressor protein shown in Fig. 1.6 provides a typical example. It is the formation of H-bonds that enables the protein to recognize its 'cognate' binding site, i.e. to read the sequence of bases on DNA. Figure 1.7 shows how a crucial glutamine residue in the (correctly positioned) λ-repressor recognition helix can form two H-bonds with an adenine base. Note that recognition does not require separation of the DNA strands.

Hydrogen bonds are relatively weak (roughly $1\,\text{kcal}\,\text{mol}^{-1}$ in water compared with $3\,\text{kcal}\,\text{mol}^{-1}$ for ionic bonds) and the few H-bonds formed by even a correctly aligned α-helix and its cognate DNA sequence will not form a stable

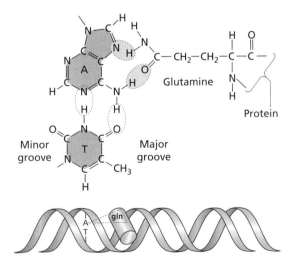

Fig. 1.7 Binding of a protein amino acid side-chain to a base in DNA. A specific glutamine residue in the DNA-binding helix of the λ-repressor protein (a helix-turn-helix protein, Fig. 1.6) is exactly positioned in the DNA major groove (lower part of the figure) so that a hydrogen atom and an oxygen atom in its side-chain are able to form hydrogen bonds with specific nitrogen and hydrogen atoms in the base adenine (upper part of the figure). These bonds are shaded to distinguish them from the ones that maintain the AT base pair. (Redrawn from Ptashne, 1992.)

complex. At best, they will cause the protein to pause as it scans along the DNA. Additional stability is provided in two ways. First, amino acids in the second, more amino-terminal, helix of the HTH–DNA binding domain interact with the DNA, sometimes with the sugar–phosphate backbone. Amino acids outside the DNA-binding domain may also be involved. These interactions are dependent upon sequence-specific recognition by the first DNA-binding helix and appropriate positioning of the protein on the DNA. Second, the binding strength of HTH proteins is significantly enhanced by the fact that they invariably bind to DNA as homodimers, i.e. pairs of identical protein subunits. Binding as a homodimer effectively doubles the free energy change that occurs as a result of binding, but causes an even greater increase in the equilibrium (or affinity) constant. (The equilibrium constant, K, and the free energy, ΔG^0, of a reaction are related by the expression $e^{-\Delta G^0/RT}$, where R is the gas constant and T is the absolute temperature.)

An initial clue that HTH proteins bind as dimers came from studying the DNA sequences that they recognize. These are usually symmetrical, consisting of two half sites with sequences that are very similar, if not exactly the same, but in opposite orientations (Fig. 1.8). For this form of binding to work, there must be strict complementarity between the tertiary structure of the protein and the structure of B-form DNA (Box 1.1). As shown in Fig. 1.9, the structure of the dimer is such that the two DNA-binding domains are precisely positioned so they can interact with adjacent major grooves.

DNA-binding proteins are made up of different functional domains

DNA binding is only one of several tasks that proteins such as *lac* repressor must be able to accomplish if they are to do their jobs properly. For example, the repressor must be able to bind lactose, to change its shape in response to this, and to bind to other *lac* repressor proteins to form dimers and possibly tetramers. Specific regions (domains) of the protein carry out these tasks. The domains of the *lac* repressor are shown in Fig. 1.10. Assignment of particular functions to particular domains is usually the result of mutation studies. For example, mutations that delete or alter the helix 3 region suppress DNA binding, whereas mutations at the C-terminal end prevent the formation of tetramers. (The precise functional role of these tetramers remains controversial.) Crucial amino acids can be identified by the functional effects of substitution mutants, and confirmed by detailed structural analysis. Domain-swaps are now a rela-

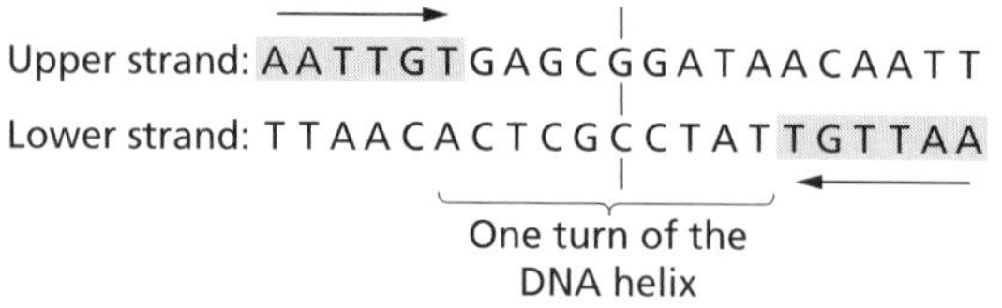

Upper strand: A A T T G T G A G C G G A T A A C A A T T

Lower strand: T T A A C A C T C G C C T A T T G T T A A

One turn of the
DNA helix

Fig. 1.8 Sequence of *lac* operator DNA.

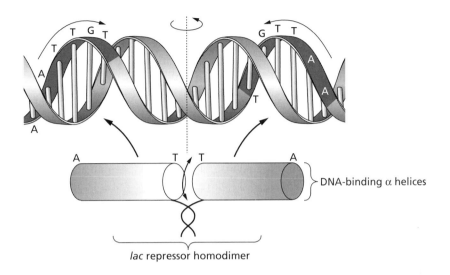

Fig. 1.9 The symmetrical base sequence of the *lac* operator allows the lac repressor protein to bind as a homodimer. The two DNA-binding helices are aligned so that they each fit exactly into adjacent major grooves of B-form DNA, where they contact the *lac* repressor sequence (AATTGT). The two helices are in opposite orientations and the dimer has an axis of rotational symmetry, shown by a dotted line running down the page. If the molecule is rotated through 180° in a plane perpendicular to the page, the same structure will reappear. The *lac* operator DNA has the same symmetry.

tively common experimental approach. For example, a DNA sequence encoding the DNA-binding domain of one protein can be spliced to a sequence encoding the ligand-binding domain of another. The novel gene, when introduced into cells and expressed, will make a hybrid protein whose properties can tell us interesting things about domain function.

DNA–protein binding can cause structural changes in both protein and DNA

The DNA molecule is sufficiently flexible for its structure to be moulded by association with proteins and it comes as no surprise that binding of HTH proteins often causes DNA bending. For example, binding of the *lac* repressor to operator DNA induces a bend *away from* the protein. This is partly a consequence of amino acids in the hinge region of the protein widening the DNA minor groove (Fig. 1.11). Conversely, binding of the CAP homodimer causes the DNA to bend *towards* the protein. DNA bending not only allows a snug fit between the protein and the DNA, but also can, in itself, be functionally important, for example by juxtaposing regions of DNA that are separated on the linear molecule and

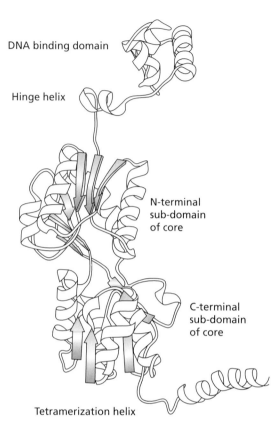

DNA binding domain

Hinge helix

N-terminal
sub-domain
of core

C-terminal
sub-domain
of core

Tetramerization helix

Fig. 1.10 The domain structure of the *lac* repressor monomer. The structure shown contains 357 amino acids. Regions of α-helix are shown as coils (some containing only one or two turns) and regions of β-sheet are indicated by flat arrows. Lines show the backbone path of residues connecting these structures. (Redrawn from Kercher *et al.*, 1997.)

thereby allowing them to interact with the same protein or protein complex. Examples of this are presented in the next chapter.

The extent of DNA bending is influenced by its base sequence. Some DNA sequences have an intrinsic curvature independent of any associated proteins. For example, runs of 3–6 A residues in one strand cause DNA to bend. It is clear from this that even bases that make no direct contact with a DNA-binding protein may influence the binding affinity through their effects on the intrinsic curvature of the DNA. An intrinsic curvature in the same direction as that induced by the protein will enhance binding affinity, whereas curvature in the opposite direction will diminish it.

DNA binding can also change the structure of the DNA-bound protein. In the *lac* repressor, the stretch of amino acids (residues 50–58) that links the DNA-binding domain and the core region forms an α-helix *only* in the presence of DNA. The helix makes specific contacts with the DNA and it seems likely that the DNA itself induces the structural rearrangement that leads to helix formation. Such transformations have been seen in a variety of different DNA-binding proteins.

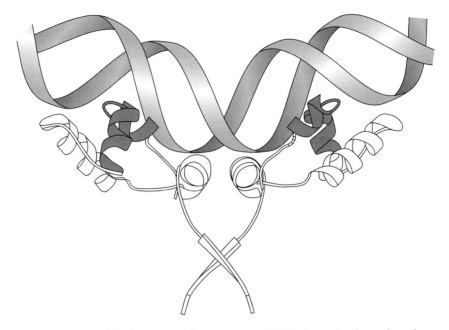

Fig. 1.11 Binding of the *lac* repressor dimer to operator DNA induces a bend away from the protein. The two DNA-binding helices are inserted into adjacent major grooves of DNA, where they form hydrogen bonds with bases of the two symmetrically aligned operator sequences. The longer α-helices close by stabilize this interaction, possibly through additional DNA contacts. The two short hinge helices contact the DNA backbone at the intervening minor groove. These multiple interactions cause the DNA to bend away from the protein. (Redrawn from Kercher *et al.*, 1997.)

Thoughts on the nature of protein–DNA binding

Two points from the last section should be emphasized strongly because of their fundamental importance for what is to follow. The first concerns how one pictures the interactions between proteins and DNA. It is customary to explain the exquisite specificity of enzymes for their substrates by a lock and key analogy: only exactly the right key will fit the lock. This is an all or nothing situation. An 'almost-right' key is essentially useless. At first sight, DNA–protein interactions operate in much the same way. The exact juxtaposition of the DNA-binding domains of HTH proteins and the major grooves of B-form DNA is certainly reminiscent of a key fitting its lock. But there are important differences. One is that both lock and key are flexible. As we have seen, both protein and DNA can undergo structural changes that serve to strengthen or weaken the interaction. Another is that an 'almost right' key can, in some circumstances, be extremely useful. Weak promoters that bind the σ-factor of RNA polymerase relatively poorly are used to give a low level of transcription of genes for which this is appropriate.

The second point that deserves emphasis is how we picture the recognition of DNA by proteins. The essential point here is that what proteins 'see' is not a sequence of bases, but a shape, a landscape of chemical entities that they may or may not be able to bind to. Sequence-specific DNA-binding proteins can track along the major groove and recognize specific base pairs, but useful, high-affinity binding cannot result if the DNA is distorted. Changes in shape will suppress binding, even though the requisite sequence of base pairs is still there.

Box 1   The basic structure of DNA

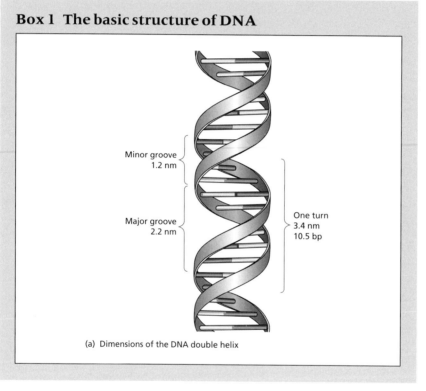

Minor groove
1.2 nm

Major groove
2.2 nm

One turn
3.4 nm
10.5 bp

(a) Dimensions of the DNA double helix

Continued

Box 1 (*continued*)

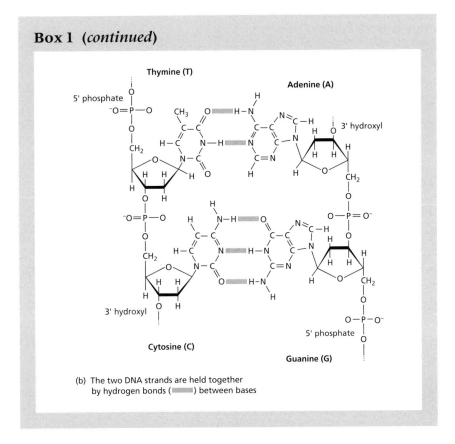

(b) The two DNA strands are held together
by hydrogen bonds (▓▓▓▓▓) between bases

Further Reading

General

Hartl, D. L. & Jones, E. W. (1998) *Genetics: principles and analysis*. Jones and Bartlett, Sudbury, MA.

Lodish, H., Baltimore, D., Berk, A. *et al.* (1995) *Molecular Cell Biology*. Scientific American Books, New York.

Ptashne, M. (1992) *A Genetic Switch*. Cell Press/ Blackwell Scientific Publications, Oxford.

White, R. J. (2000) *Gene Transcription*. Blackwell Science Ltd, Oxford.

The RNA polymerase–promoter complex

Naryshkin, N., Revyakin, A., Kim, Y., Mekler, V. & Ebright, R. H. (2000) Structural organization of the RNA polymerase–promoter open complex. *Cell*, **101**: 601–611.

McClure, W. R. (1985) Mechanisms and control of transcription initiation in prokaryotes. *Ann. Rev. Biochem.*, **54**: 171–204.

Protein–DNA binding

Harrison, S. C. (1991) A structural taxonomy of DNA binding domains. *Nature*, **353**: 715–719.

Kerchner, M. A., Lu, P. & Lewis, M. (1997) *Lac* repressor–operator complex. *Curr. Opin. Struct. Biol.*, **7**: 76–85.

Pabo, C. O. & Sauer, R. T. (1992) Transcription factors: structural families and principles of DNA recognition. *Ann. Rev. Biochem.*, **61**: 1053–1095.

Parkinson, G., Wilson, C., Gunasekera, A. *et al.* (1996) Structure of the CAP–DNA complex at 2.5 Å resolution: a complete picture of the protein–DNA interface. *J. Mol. Biol.*, **260**: 395–408.

Protein–protein interactions

McNight, S. L. (1991) Molecular zippers in gene recognition. *Sci. Am.*, April 1991: 32–39.

Chapter 2: Transcription in Eukaryotes: The Problems of Complexity

Introduction

In the previous chapter we used information about gene expression in relatively simple, prokaryotic organisms to introduce some important concepts about how genes can be switched on and off. These model systems are particularly valuable because the molecular machinery that drives this most fundamental of cellular processes has many features that are common to organisms at all levels of complexity. However, it is also an inescapable fact that as organisms become more complex and the information needed for their assembly and day-to-day management increases, so their genomes grow bigger. As we shall see shortly, an increase in genome size brings both opportunities and problems. The control systems that work so well for small genomes can be overwhelmed by the amount of genetic information needed at higher levels of complexity. To cope with this, the old systems have not been discarded but have been added to and modified.

The emergence of eukaryotes

The best estimates we have suggest that the first single-celled organisms appeared on Earth about 3.5 billion years ago. About 2 billion years later, some single-celled organisms developed a new way of packaging their genetic material by sequestering it into a membrane-bound subcellular organelle, the nucleus, and associating it with a particular family of proteins, the histones. They also acquired both cytoplasmic membranes and organelles such as mitochondria. These innovations gave rise to a new family of single-celled organisms, the eukaryotes. How these fundamental changes in cellular structure and organization came about is not clear, but the enormous length of time between the first appearances of prokaryotic and eukaryotic cell types suggests that the evolutionary steps involved were both complex and intrinsically unlikely. They may have involved fusion of cells so as to form a new (more efficient) cell type, perhaps as a consequence of a symbiotic relationship between two different prokaryotic cells. Certainly there is strong circumstantial evidence that mitochondria, which still retain their own bacterial-type genomes, first arose in this way.

Whatever its origins, it is hard to overstate the importance of this evolutionary step. Only eukaryotes have been able to evolve still further and develop the additional genetic complexity needed to form multicellular organisms. The possession of a nucleus, and a rather particular method of packaging DNA within it, together with cytoplasmic organelles, allowed not only more efficient energy generation and metabolic control but also, and crucially, provided additional possibilities for regulating the expression of a larger number of genes. We will return to this at the end of the chapter.

The simplest eukaryotes

Not all eukaryotes have gone on to higher things. Some, such as the yeasts, have remained single-celled. Despite (or because of) their simplicity, these single-celled eukaryotes are particularly valuable to human beings. Not only are they essential for baking and brewing, but they also provide almost ideal experimental organisms. They are easy (and cheap) to grow under laboratory conditions, they grow rapidly (cell numbers double in 1–2 h) and, most importantly, they are amenable to genetic manipulation. Specific genes can now be inserted, deleted, repositioned or changed to provide a genetic tool-kit of unparalleled power for functional analysis of individual genes and their protein products. Data derived from experiments with the two (evolutionarily distant) yeasts *Saccharomyces cerevisiae* and *Schizosaccharomyces pombe* features prominently in subsequent chapters.

Despite being only single-celled, *S. cerevisiae* has many more genes than the prokaryote *E. coli* (Table 2.1). In fact, *E. coli* is a relatively complex prokaryote, so the difference, in general terms, between single-celled eukaryotes and prokaryotes is even greater than the examples shown in the Table. The extra genetic information in single-celled eukaryotes is used, amongst other things, to construct their additional organelles, to encode the enzymes responsible for additional metabolic pathways (such as oxidative phosphorylation) and to allow biological behaviours such as mating (yes, even yeast do this) and meiosis.

Table 2.1 Examples of genome size and gene density.

Organism	DNA (Mb) (haploid)	Genes (protein coding)	Chromosomes	Gene density (genes/Mb)
Bacterium (*E. coli*)	4.5	4000	1	900
Yeast (*S. cerevisiae*)	15	6200	17	400
Worm (*C. elegans*)	100	18400	6	180
Fruitfly (*D. melanogaster*)	120*	13600	4	110*
Human (*H. sapiens*)	3300	~30000	23	9

*This organism also has 60 Mb of largely repetitive, non-coding DNA ('heterochromatic' DNA) that is not included in the calculation (see Chapter 9). The same type of DNA is found in humans and other mammals, but in amounts that do not significantly affect the calculation of gene density.

Multicellular organisms

A second major evolutionary step, apparently possible only for eukaryotes, is the development of a body form consisting of several *different* cell types, i.e. multicellularity. It is rather easier to envisage how this might have occurred than how the transition from prokaryotic to eukaryotic cell types was made. Both prokaryotic and eukaryotic organisms can form loose clusters of cells under certain environmental conditions and a simple eukaryote, the slime mould *Dictiostelium discoidium*, has a life cycle that contains both unicellular and multicellular stages, the latter containing several very different cell types. Certainly multicellular eukaryotes evolved from their single-celled ancestors in only a fraction of the time that it took for eukaryotic cells to evolve from prokaryotes, probably about 500 million years.

Multicellularity is expected to bring with it a need for increased genetic information and hence a larger genome. First, the genome must encode the information necessary to construct and operate several different cell types. Second, it must encode instructions necessary to regulate this information appropriately, i.e. to ensure (amongst other things) that genes that are needed for a particular cell type are switched on in that cell type and off in all others. Third, it must encode the information necessary to enable the different cell types to communicate with one another. The pattern of gene expression in a particular cell is often influenced strongly by the cells that surround it, either through physical contact or through soluble signalling molecules that diffuse through the extracellular milieu. Thus, a multicellular organism must have the genetic information to enable it to both receive and transmit a much larger variety of signals than are needed by unicellular life forms. All this leads one to expect that multicellular eukaryotes will have more complex (larger) genomes than unicellular ones, and the figures bear this out.

The slime mould *D. discoidium*, which is multicellular only at certain stages of its life cycle, has about 12.5 thousand genes while the fruitfly *Drosophila melanogaster* (another favourite experimental organism), has about 13.5 thousand. As complexity of body form increases so does gene number. Thus, the nematode worm *Caenorhabditis elegans* has fewer than 30 different cell types in the adult and 18 500 genes whereas humans, with just over 200 different cell types, have about 30 000 genes. (This is based on the results of the project, now virtually complete, to sequence the human genome. It is 2–3 times lower than previous estimates and is still not exact, primarily because it is often difficult to tell whether a given sequence of DNA constitutes a functional gene or not.) At this stage, it is worth introducing a note of caution about 'complexity' in the context of the present discussion. As humans, we like to think of ourselves as more complex than other vertebrates, such as mice or fish, and in some ways we are. But in biological terms, i.e. in terms of the numbers of different cell types and of the developmental problems that must be solved in forming and assembling them, we are not so very different. In fact, the mouse (*Mus musculus*) and the puffer fish (*Fugu rubripes*) both have about the same number of genes as ourselves.

Gene frequency decreases as genomes get bigger

Table 2.1 summarizes the figures so far available for DNA content and genome size in some example organisms. Not only do gene numbers increase as organisms get more complex, but gene density (i.e. genes per Mb) falls dramatically. For example, there are 890 genes per Mb in *E. coli* but only 9 genes per Mb in humans. Why is this so? The major reason is the presence in eukaryotes of significant amounts of noncoding DNA. This is sometimes called 'junk' DNA, the implication being that it has accumulated over evolutionary time through essentially random insertions (perhaps of viral DNA), duplications or progressive loss of coding capacity of redundant genes. All of these things certainly occur, but the term is misleading because it implies lack of function. As we will see later, noncoding DNA plays an important role in the correct positioning of protein binding sites. This may be on a small scale so as to facilitate interactions between two proteins binding to adjacent sites, or on a larger scale requiring folding of several kilobases of intervening DNA. On the largest scale, long stretches (many Mb) of noncoding, often repetitive, DNA can form specific DNA–protein structures (heterochromatin) that can exert a dominant influence on the behaviour of genes in their vicinity. Indeed, the extraordinary amount of repetitive, noncoding DNA in *Drosophila* (about one-third of the total DNA; Table 2.1) can exert an overwhelming effect on transcription (Chapter 6). We certainly can't say that repetitive DNA has evolved as it has *because* of its ability to influence transcription, but it is something we would be unwise to ignore.

Genes in higher eukaryotes contain stretches of noncoding DNA

In higher eukaryotes, noncoding DNA is found not only between genes but also within genes themselves. These sequences, known as *introns*, are removed in a complex process called splicing that is also part of the mechanism by which mRNAs are transported from the nucleus to the cytoplasm. The splicing machinery can, in some circumstances, cause several different protein products to be produced from the same sequence of DNA through the use of different splice sites, so-called *alternative splicing* (Fig. 2.1). Given the complexity of the splicing machinery, this seems a rather small benefit. The evolutionary history of introns and splicing remains unclear, but it may be that splicing arose as a by-product of the development of systems for transporting RNA transcripts from nucleus to cytoplasm, where they are translated. Introns account for only a small part of the increase in noncoding DNA in higher eukaryotes compared with prokaryotes.

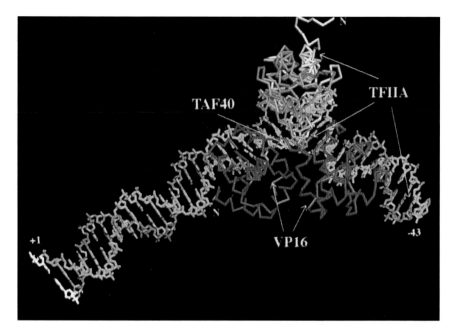

Plate 2.1 Stick model showing the TATA-binding-protein (TBP) and the transcription factor TFIIB bound to the DNA of the adenovirus major late promoter (AdMLP). The structure is derived from X-ray diffraction studies of TBP–TFIIB–AdMLP crystals. TBP is shown in light/dark blue and TFIIB as orange/magenta. N-termini are indicated (N). Complementary DNA strands are shown as yellow and green. The transcription start site is indicated (+1, transcription would proceed to the left of the figure). Regions of TBIIB implicated in binding two other components of the transcription initiation complex, $TAF_{II}40$ and TFIIA, are indicated, together with the region that binds to the activator VP16. Binding of TBP–TFIIB bends the promoter DNA through about 110°. (From Nikolov, D.B. *et al.* 1995, *Nature* **377**, 119.)

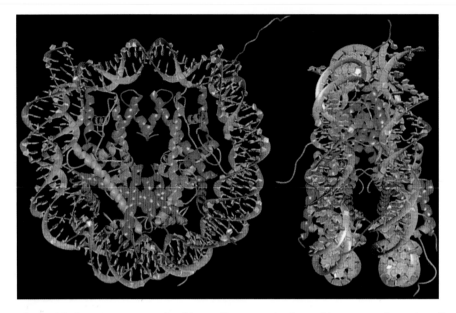

Plate 3.1 Computer-generated ribbon diagram showing the structure of the nucleosome core particle based on X-ray crystallographic analysis at 2.8Å resolution. The paths of the sugar–phosphate backbones of the complementary strands of the DNA helix are traced in brown and green with the base pairs shown as lines of corresponding colour. The main chains of the 8 core histones are shown in yellow (H2A), red (H2B), blue (H3) and green (H4). Coiled regions represent α-helices. The left-hand view is looking down the axis of the DNA superhelix and the right hand view is at right angles to this axis. The dyad axis (Fig. 3.6a) is at the top. (From Luger *et al*. 1997.)

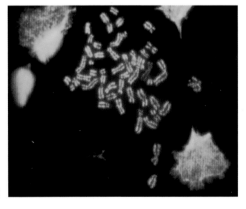

(a)

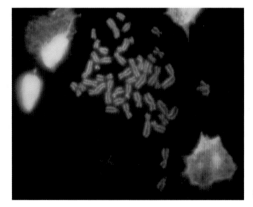

(b)

Plate 5.1 (a) Metaphase chromosomes from a human female lymphocyte spread on a glass slide and immunostained with an antibody to acetylated histone H4. The antibody is visualized with a green fluorescent dye (FITC). Regions that fluoresce strongly are rich in acetylated H4 and regions that show little or no fluorescence are depleted in acetylated H4. Note that heterochromatin at and around the centromere is usually underacetylated (the chromosome at the top centre provides a good example). Many chromosomes show a clear pattern of bright and dim bands along the chromosome arms. The bright (i.e. acetylated) regions may correspond to gene-rich R-bands (Chapter 5). One chromosome is weakly stained overall. This is the transcriptionally silent, inactive X chromosome (Xi) found only in female cells (see Chapter 12). Most of the genes on this chromosome are transcriptionally inactive. An exception is a small region at the end of the short arm, the pseudoautosomal region, in which genes remain active. Histones in this region remain acetylated. (b) The same chromosome spread as in (a) stained with the DNA-binding fluorochrome Hoechst 33342.

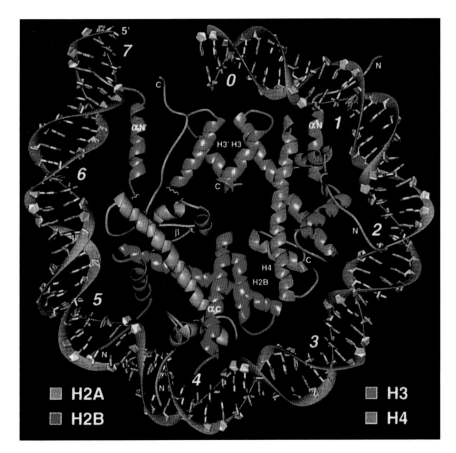

Plate 3.2 Computer-generated ribbon diagram showing the path of the single turn of the DNA helix around the histone core and histone–histone interactions within the core. One complete molecule of each histone is shown, together with part of the second H3 molecule. (H2A, yellow; H2B, red; H3, blue; H4 green). The 'histone handshake' motif, by which H2A interacts with H2B, and H3 with H4 is clearly seen (see also Fig. 3.7). Those regions of the histone N-terminal and C-terminal tail domains that remain inside the DNA helix are indicated by N and C respectively. Histones H2B and H3 each have an extra region of α-helix outside the histone fold (Fig. 3.7). These regions are labelled αC (H2B) and αN (H3). An interface between H2A and the C-terminal region of the second H4 molecule is labelled β. Complete turns of the DNA helix are numbered 0 to 7, with 0 being positioned at the dyad axis (these turns are shown diagrammatically in Fig. 4.2). Note the variable curvature of the DNA as it folds around the histone core. (From Luger *et al.* 1997.)

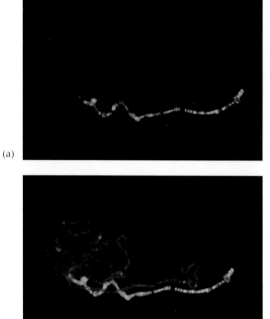

(a)

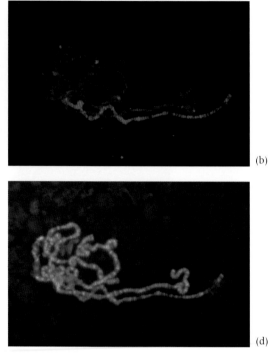

(b)

(c)

(d)

Plate 12.1 (a) Polytene chromosomes from the fruit fly *Drosophila melanogaster* immunostained with an antibody to the MLE protein, visualized with a green fluorochrome (FITC). MLE is a component of the dosage compensation complex that binds to specific sites along the X chromosome in male flies and leads to a two-fold up-regulation in transcriptional activity (see Chapter 12). (b) The same chromosome spread was stained also with an antibody that specifically recognises histone H4 acetylated at lysine residue 16. This antibody was visualized with a red fluorochrome. Note that similarity to the staining pattern in (a).

(c) Simultaneous visualization of the red and green fluorochromes shows that they are usually, though not always, located at the same positions along the chromosome, generating a yellow colour. (d) The same chromosome spread stained with the DNA-binding fluorochrome DAPI. The brightly staining bands represent regions of relatively compacted DNA/chromatin revealed by various DNA stains, by phase contrast microscopy or by electron microscopy (e.g. Figs 5.17, 5.18, 6.3). (Photographs provided by Mitzi Kuroda; see Bone *et al.* 1994.)

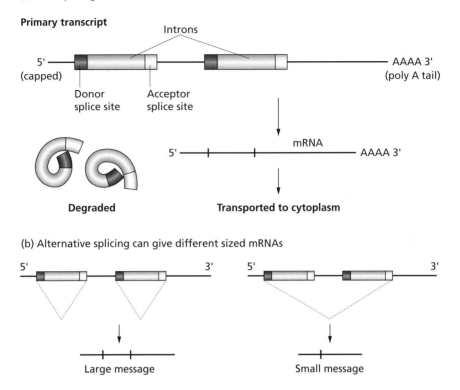

(a) RNA splicing removes introns

Primary transcript

Introns

5' ——— (capped)   Donor splice site   Acceptor splice site   AAAA 3' (poly A tail)

Degraded   mRNA   5' ——————— AAAA 3'   **Transported to cytoplasm**

(b) Alternative splicing can give different sized mRNAs

5' ——— 3'   5' ——— 3'

Large message   Small message

Fig. 2.1 RNA splicing removes introns. Most eukaryotic genes contain introns, stretches of DNA that are transcribed along with the rest of the gene, but that do not form part of the coding sequence. Introns are removed from the RNA transcript by a complex set of enzymes located at specific sites in the nucleus, and then degraded. The spliced RNA transcript is transported to the cytoplasm for translation.

The transcription machinery in eukaryotes

In the second part of this chapter we will discuss the molecular machinery of transcription in eukaryotes. The similarities with their prokaryotic equivalents are striking evidence of their common evolutionary origins, whereas the differences and added complexities can often be understood in terms of the demands imposed by a larger genome.

Eukaryotic RNA polymerases

Three different RNA polymerases have been identified in eukaryotic cells. They are referred to as PolI, PolII and PolIII and their subunit compositions are listed in Table 2.2. The data in the table refer to the enzymes from *S. cerevisiae*, but, in fact, the properties of these enzymes are very similar from one eukaryote to

Table 2.2 Eukaryotic RNA polymerases.

Polymerase	Genes transcribed	Core subunits		Shared subunits	Specific subunits
		β, β'-like	α-like		
PolI	rRNA	135+190 kDa	19+40 kDa*	Six (10–27 kDa)	Five (12–49 kDa)
PolII	protein-coding	150+220 kDa	44 kDa (two)	Six (10–27 kDa)	Four (12–32 kDa)
PolIII	small RNAs (e.g. tRNAs)	128+168 kDa	19+40 kDa*	Six (10–27 kDa)	Seven (11–82 kDa)

*These subunits are the same in PolI and PolIII.

another. There are obvious parallels between the subunit structures of the three enzymes and the single, major RNA polymerase found in prokaryotes such as *E. coli*. All three eukaryotic enzymes have subunits that resemble the bacterial β, β' and α subunits in size, number per enzyme molecule and amino acid sequence. There are clear homologies in the amino acid sequences of β' subunits from yeast, *Drosophila* and *E. coli*. These observations point to the common evolutionary origin of the prokaryotic and eukaryotic enzymes.

The eukaryotic polymerases are clearly more complex structures than the prokaryotic enzyme. While the latter has only one subunit in addition to α, β and β' (i.e. the σ subunit that confers promoter specificity and DNA binding), the eukaryotic enzymes each have between 10 and 13 additional subunits. Six of these are present in all three enzymes and the rest are specific for a particular polymerase (Table 2.2). The major reason for the three different enzymes in eukaryotes seems to be that they are specialized for the transcription of particular sets of genes. PolI transcribes the genes that encode the major ribosomal RNAs, PolII transcribes the genes encoding the great majority of the cell's proteins, and PolIII transcribes genes encoding small RNAs such as transfer RNA (tRNA) and small (5S) ribosomal RNAs. The genes encoding the various types of structural and functional RNAs transcribed by Pols I and III are all present in multiple copies (as they are in *E. coli*) in order to meet the demands of the cell for these components. Remember that for these genes, the RNAs produced are themselves the functional product; there is no further product amplification through the process of translation, as occurs with protein-coding genes. The different structures of PolI, PolII and PolIII have presumably evolved to accommodate differences in the organization and the required rates of transcription of the genes they transcribe, which, in turn, have been moulded to meet the requirements of the cell for the gene product. It is interesting to note that a hint of what was to come (in evolutionary terms) can be seen in prokaryotes in the fact that the structure of the predominant RNA polymerase can be adjusted to facilitate transcription of specific genes. This is achieved either by association with minor σ factors whose DNA-binding specificities allow targeting to specific promoters, or by replacement of the usual α subunits with slightly different ones in order to facilitate transcription of rRNA genes, of which there are seven copies in *E. coli*.

Of the three eukaryotic polymerases, we will be concerned mostly with PolII. This is the enzyme responsible for transcription of protein-coding genes and it is the presence or absence of transcription of selected members of this group of genes that determines the fate of the cell through differentiation and development. The problems of gene regulation with which we are concerned here are essentially problems of control of PolII. However, PolI and PolIII are not only essential for cell survival, but also are sometimes used as experimental models from which attempts are made to draw general conclusions. We will return to them towards the end of the chapter.

Eukaryotic promoters

In *E. coli*, promoters for protein-coding genes have a common, bi-partite structure within which the actual DNA base sequence varies widely, giving a range of promoters that show either strong or weak binding of RNA polymerase. This variability is a useful element in the control of transcription rates. In eukaryotes, promoters for protein-coding (i.e. PolII-transcribed) genes vary even more widely. Three DNA sequence elements are present in eukaryotic promoters, either singly or in combination (Table 2.3). These are the TATA box (so named because of its defining TATA consensus sequence), the initiator element and elements rich in the dinucleotide CpG. The last are known as 'CpG islands' and will be discussed in much more detail in Chapter 9. Genes whose transcription varies widely, either from one cell type to another or through the cell cycle, tend to have promoters with a TATA box.

Eukaryotic RNA polymerases cannot bind to promoters, or to DNA in general, on their own. They can be targeted to the promoter only through association with other protein factors. These other proteins provide both promoter-specific DNA binding and the capacity to recognize RNA polymerase and thereby position it correctly at the promoter. (Their role can be compared to that of CAP and other activator proteins in positioning polymerase at promoters in *E. coli*.) It is

Table 2.3 DNA elements in eukaryotic promoters.

Element	Defining sequence elements	Distance from start site	Genes in which this type of promoter tends to be found
TATA box	. . . TATA$^A/_T$A. . . .	−20 to −35	Rapidly transcribed, often cell-cycle or tissue-specific (e.g. histones, globins)
Initiator	. . . PyCAN$^T/_A$PyPy* . . .	CA at −1, +1	Various (frequent in viral promoters)
CpG island	CpG-rich region of 20–50 bp	−100 to −200	Slowly transcribed (e.g. enzymes of intermediary metabolism)

*N = any nucleotide, Py = a pyrimidine base (C or T)

important to note that the promoter is only a positioning sequence. The DNA sequence elements that are commonly found in promoters are not of any great significance for the process of transcription itself. RNA polymerase will transcribe any DNA sequence to which it can be targeted, provided it has access to the necessary chemicals and auxiliary proteins.

A typical eukaryotic gene

For most of this book we will be concerned with the protein coding genes transcribed by PolII. No single gene can reasonably be presented as representing PolII genes in general, but the hypothetical gene shown in Fig. 2.2 incorporates the basic DNA regions involved in the regulation of all PolII genes.

Upstream DNA elements that bind specific proteins (transcription factors) are a feature of PolII genes. Most of these elements are 'promoter proximal' (i.e. are within 200 base pairs (bp) or so of the TATA box) but some 'enhancer' sites are located several kilobases (kb) from the promoter. These enhancers are often upstream (i.e. 5′) of the promoter, but can be downstream (e.g. the *Igf2/H19* genes discussed in Chapter 10) or even in an intron within the coding region. The proteins that bind to these sites often interact, directly *or via intermediary proteins*, with the RNA polymerase complex. The function of these proteins is twofold. First, they organize the DNA in such a way that it is structurally able to accommodate the RNA polymerase complex. They must ensure that the promoter is accessible and that other proteins are positioned so as to interact in a useful and productive way. They do this in conjunction with the histones and other proteins that package DNA as chromatin. It is this interaction that determines whether, and at what rate, a functional pre-initiation complex (PIC) can be assembled at any given promoter. The second role of these DNA-binding proteins is to stabilize the polymerase and associated proteins once they arrive, and thereby allow assembly of the complete, functional PIC. They do this by interacting with specific protein components of the polymerase complex itself.

These DNA-binding proteins contain exactly the same amino acid sequence motifs and tertiary structures as were described for DNA-binding proteins from prokaryotes in Chapter 1 (i.e. helix-turn-helix, etc.). They also have domains that are required for the very specific protein–protein interactions that enable

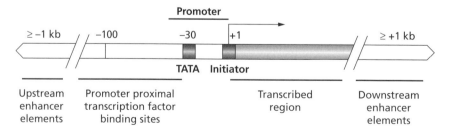

Fig. 2.2 Distribution of regulatory elements in a eukaryotic gene.

them to facilitate PIC assembly. They will often act in a synergistic fashion, i.e. two proteins together will exert an effect that far exceeds their effects individually. Some of the promoter proximal sites are present upstream of many promoters and the transcription factors that bind them are therefore employed by many different genes (proteins such as NF1 and Sp1). Others are restricted to a limited number of genes and often mediate responses to a defined chemical or environmental factor. They include the cyclic AMP response element (CRE) and the heat shock element (HSE). More than one copy of a particular DNA element may be found upstream of a single promoter.

In order to function properly, proteins bound to upstream sites must be able to interact with the PIC assembled on and around the TATA box. The binding sites for these proteins can be over 200 bp upstream of the TATA box. This represents 70 nm of B-form DNA, too great a distance to allow protein–protein interactions without some form of DNA bending. Bending is also necessary to allow *multiple* upstream proteins to interact with the PIC. The manipulation of the DNA helix to facilitate these interactions is the job of the so-called 'architectural' proteins (e.g. nonhistone chromatin proteins such as the ubiquitous HMG proteins) and of chromatin itself (Fig. 2.3).

The need to manipulate DNA becomes even more acute when considering the mechanism by which sequences far upstream of the promoter can influence

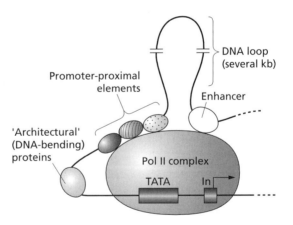

Fig. 2.3 Proteins bound to DNA elements upstream of the promoter interact with components of the RNA PolII complex. The PolII complex assembles at the promoter (more details are given in Fig. 2.4). Many, although not all, eukaryotic promoters contain a TATA sequence and an 'initiator' sequence at the transcription start site (Fig. 2.2, Table 2.3). Proteins bound to DNA elements upstream (usually) of the transcription start site facilitate assembly of the complex or help it to function appropriately. Interaction with the complex, of proteins bound to DNA close to the promoter, can be assisted by proteins that bend DNA. 'Enhancer' elements can be located several kilobases from the transcription start site. The intervening DNA must be folded to allow proteins bound to enhancer elements to act upon the PolII complex. This process is often described as 'looping' of the DNA, but the actual structures involved are unknown.

transcription. Such sequences, generally referred to as 'enhancers' can be several kb upstream of the promoter they regulate. Transcription of some genes is absolutely dependent on the action of such enhancers. In some cases, these upstream regions are involved in more complex regulatory problems. For example, the locus control region (LCR) 50 kb upstream of the human β-globin locus functions not only as an enhancer, but is also necessary for the correct expression through development of other genes in the β-globin gene cluster. The inherited absence of part of this region in humans results in a form of thalassaemia (haemoglobin deficiency and severe anaemia), even though the globin genes themselves and their promoters are perfectly normal. Bringing the LCR into proximity of the promoters of β-globin and, when required, of other genes in the cluster requires some extensive DNA manipulation. This is usually referred to as 'looping' and is shown, in schematic form, in Fig. 2.3, but the truth is that we do not yet have any clear idea of how the larger-scale (re)positioning of DNA sequences within the nucleus is achieved. This is an issue we will return to repeatedly.

General transcription factors, TAFs and the PolII pre-initiation complex

Figure 2.3 defines the general principles underlying PIC assembly. The next section fills in some important details about the composition of the PIC and the route by which it is assembled. Much remains to be done in this area and the details are not often well understood, but the major players in the process have been identified and a plausible assembly pathway has been proposed. This is shown in Fig. 2.4.

The PIC consists of the PolII complex together with a set of six *general transcription factors* designated TFIIA, TFIIB, etc., all of which are required for initiation of transcription. Each of these is, itself, a multiprotein complex. Some details of the various components of the human transcription initiation complex are listed in Table 2.4. Essentially the same components have been identified in all eukaryotes so far studied, although their detailed compositions may vary. (The nomenclature of these complexes is, fortunately, fairly logical. TF stands for Transcription Factor, II indicates that we are dealing with PolII, and the letters A to H indicate the order in which the complexes were identified and characterized; C and G seem to have been lost along the way.) The first step in PIC assembly involves binding, not of the polymerase itself, but of the transcription factor TFIID, along with the much smaller complex, TFIIA. TFIID has at least 12 different protein subunits, one of which is the TATA-binding protein (TBP). TBP is required for recruitment of RNA polymerase to the great majority of eukaryotic promoters, even those that do not contain a TATA box. This becomes less puzzling when one realizes that, although TBP has a high affinity for DNA, its specificity for the TATA box is much lower than that of a typical, sequence-specific DNA-binding protein. In fact, a TATA box alone does not

Table 2.4 Components of the human transcription initiation complex (in order of their recruitment).

Factor	Subunits	Size range (kDa)	Some functions
TFIID—TBP	1	38	TATA box recognition, TFIIB recruitment
TFIID—TAFs	12	15–250	Promoter (non-TATA) recognition, regulatory functions (see Table 2.5)
TFIIA	3	12, 19, 35	Stabilization of TBP/TAF–DNA interactions
TFIIB	1	35	Recruitment of PolII and TFIIF
TFIIF	2	30, 74	Promoter targeting of PolII, reduces *non-specific* PolII–DNA interactions
RNA PolII	12	10–220	Catalytic function in RNA synthesis, recruitment of TFIIE
TFIIE	2	34, 57	Recruitment of TFIIH and modulation of its catalytic activities
TFIIH	9	35–89	Helicase activity separates DNA strands at the promoter; kinase activity helps initiate elongation

Taken from Roeder, R.G. *Trends Biochem. Sci.*, **21**: 327–335.

function as a promoter *in vivo*. So, although positioning of TBP on DNA is a crucial step in assembly of the PIC, the TATA box itself is only one element in this process and, in some promoters, its function can be taken over by other DNA sequences.

TFIID consists of TBP plus a set of additional proteins designated TAFs (TBP-associated factors). These proteins, and the composition of TFIID itself, have been highly conserved through evolution, with comparable complexes existing in humans, flies and yeast. The human TAFs known at present, together with some information on their properties, are listed in Table 2.5. The various TAFs play different roles in transcription initiation, and these may vary in importance from one promoter to another. It is possible that the composition of TFIID varies, depending on the specific needs of the promoter to which it is recruited. The various functions of the TAFs are considered below under three broad headings.

TAFs bind DNA sequences outside the TATA box to provide promoter specificity

TFIID binding can protect DNA between 50 kb upstream and 35 kb downstream of the TATA box from digestion by added nucleases, but the extent of protection (and thus of TFIID binding) varies from one promoter to another. In *Drosophila*, $TAF_{II}150$ is able to bind to the region downstream of the TATA box, including the 'Initiator' region at the transcription start site. The combined actions of $TAF_{II}150$ and the largest TAF, $TAF_{II}250$, seem to be necessary for initiator-dependent transcriptional activation. As noted above, the TATA box alone cannot initiate transcription *in vivo*. Binding to adjacent DNA sequences via the TAFs both adds to the specificity of binding, by recognizing specific sequences adjacent to the TATA box itself, and stabilizes the association of TBP with the promoter region.

Table 2.5 Some components of human transcription factor TFIID.

Component*	Structural features	Contacts with other proteins	Other functions
TBP	Small (30 kDa), unusual DNA-binding properties via *minor* groove	Various, mostly via N-terminal tail	The *only* DNA-binding component, bends DNA by widening minor groove
TAF$_{II}$250	HMG box, bromodomains		Acetyltransferase and protein kinase activities
TAF$_{II}$135		TFIIA	
TAF$_{II}$80	resembles H4, histone fold	TFIIE, TFIIF	Part of histone octamer-like structure?
TAF$_{II}$55		multiple activator proteins (E1A, YY1)	
TAF$_{II}$31	resembles H3, histone fold	TFIIB, acidic activator proteins (p53)	Part of histone octamer-like structure?
TAF$_{II}$30		Estrogen receptor	
TAF$_{II}$20	resembles H2B		Part of histone octamer-like structure?

*TBP, **T**ATA **B**inding **P**rotein; TAF$_{II}$, TBP **A**ssociated **F**actor (Pol**II**), followed by its molecular mass.

TAFs bind to 'activator' and 'coactivator' proteins that facilitate the promoter binding step

The binding of TBP to the TATA box is the slowest (and therefore the rate-limiting) step in transcription initiation. Anything that speeds up (or slows down even further) this critical step will have a major effect on the rate of transcription. For many eukaryotic genes, DNA sequences a short way upstream of the TATA box are binding sites for proteins that facilitate TBP recruitment and thereby activate transcription. Some of these transcription factors are found adjacent to many different promoters (e.g. Sp1) while others are more restricted (e.g. Estrogen Receptor). A few of these activator proteins can bind to TBP itself, but most bind to specific TAFs and thereby enhance recruitment of the TFIID complex. For example, Sp1 interacts with TAF$_{II}$110 and the Estrogen Receptor with TAF$_{II}$30. In some cases, TAFs (or TBP itself) interact with DNA-binding activator proteins not directly, but through intermediaries called 'coactivators'.

TAFs help assemble general transcription factors into the complete pre-initiation complex (PIC)

Binding TFIID to promoter DNA initiates the assembly of the PolII PIC and recruits other general transcription factors. TAFs are involved in the assembly of these complexes through specific protein–protein interactions. The events involved in assembly of the PIC are shown, diagrammatically, in Fig. 2.4. The diagram shows the sequential addition of the various TFII complexes, culminat-

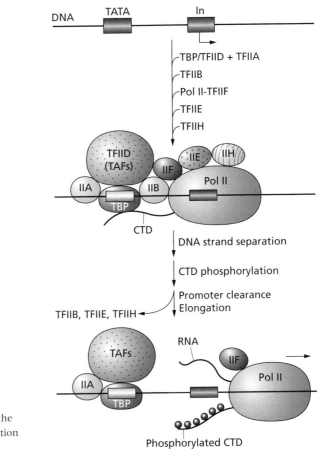

Fig. 2.4 Assembly of the transcription preinitiation complex.

ing in the complete initiation complex. There is some debate about whether this is what actually occurs *in vivo*, or whether a PolII 'holoenzyme' consisting of PolII plus TFIIB, TFIIF, TFIIE and TFIIH is pre-assembled and recruited as a complete complex. This may occur in yeast. But, the important points at this stage are that the initial recruitment of TFIID (aided by TFIIA) leads to the assembly on the promoter of a complete, PolII-containing PIC. The functions of the various TFII complexes, or their compositions, are still not completely understood. However, some significant properties have been associated with individual members of the family (Table 2.4).

Crystallographic analysis of the TBP–TATA box complex has shown that TBP binds to DNA by inserting a wedge of hydrophobic amino acids into the minor groove. This unusual binding mechanism severely distorts the DNA, causing both untwisting of the double helix and bending of the DNA helical axis. TBP binding provides an example (albeit on a small scale) of the type of DNA manipulation shown diagrammatically in Fig. 2.3. It has been possible also to crystallize a ternary complex containing DNA, TBP and TFIIB, the next transcription factor to attach (Fig. 2.4). This has allowed a detailed analysis of the protein–protein interactions that bring about this assembly step. A computer generated image of this complex is shown in Plate 2.1.

Transcription by PolI and PolIII

PolI transcribes the multicopy ribosomal RNA (rRNA) genes while PolIII transcribes the genes encoding tRNAs and other small nuclear RNAs. The promoters for the great majority of these genes lack TATA boxes, but their transcription still requires TBP. Presumably it is the ability of TBP to recruit other proteins that is important for transcription of these genes, rather than its DNA-binding ability. In fact, mutations to TBP that prevent its binding to the TATA box do not inhibit PolI and PolIII transcription. Initiation of trancription by PolI requires binding of a two-subunit protein, upstream binding factor (UBF) to a GC-rich site just upstream of the transcription start site. UBF recruits a complex containing TBP and three TAFs, $TAF_{III}48$, $TAF_{III}63$ and $TAF_{III}110$. This complex in turn recruits PolI. Binding of one or more of the TAF_{III} proteins inhibits the ability of TBP to bind to the TATA box, thereby preventing the complex from binding to PolII promoters.

The PolIII story is slightly more complex in that genes transcribed by PolIII have at least three different types of promoter. Transcription of the tRNA genes requires two 10 bp elements that lie *within the coding region* of the gene. These two elements bind a multisubunit transcription factor (TFIIIC) that in turn recruits and positions PolIII. The genes encoding the small ribosomal RNA (5S RNA) are present in about 2000, tandemly repeated copies in human cells and are also transcribed by PolIII. Like the tRNA genes, the 5S RNA genes also have a crucial control region located within the coding region of the gene, but in this case the region binds TFIIIA, a 40 kDa polypeptide with several DNA-binding zinc fingers. TFIIIA recruits the factor TFIIIC, which then, as for the tRNA genes, recruits PolIII. A third type of PolIII promoter is found in the U6snRNA genes that encode the small nuclear (sn)RNA components of ribonucleoproteins (RNPs) involved in RNA processing. These genes lack internal control elements, but do contain a correctly positioned TATA box motif.

Although the PolI and PolIII PICs lack the complexity of their PolII counterparts, at least in terms of the sheer number of components, there are familiar themes. For all three polymerases, recruitment of the enzyme itself to the promoter involves, as a first step, association of a sequence-specific DNA-binding protein with one or more DNA elements. Other proteins then attach to the DNA-binding protein (in the case of PolII they are already attached as part of the TFIID complex) and, in turn, both recruit and position the appropriate polymerase.

The elongation stage

The assembly of the PIC represents the first phase in the initiation of transcription. It is the slowest step and the one that is most susceptible to effective regulation. However, the steps that follow can also play important regulatory roles. The first of these, as in prokaryotes, is the energy-dependent separation of the two strands of the DNA double helix adjacent to the transcription start site (i.e. switching from the 'closed' to the 'open' complex). Next, the polymerase itself is released from the PIC. This is brought about, in part, by phosphorylation of the C-terminal domain (CTD) of the β-subunit of PolII by a protein kinase activity present in TFIIH. PolII is thought to be tethered to the PIC by its CTD and phosphorylation weakens this tether and allows the two components to separate. At this stage, TFIIB, TFIIE and TFIIH seem to dissociate from the complex, TFIID and TFIIA remain attached to the promoter while the polymerase itself, plus TFIIF begins transcribing (Fig. 2.4). This is reminiscent of the situation in prokaryotes where the σ factor that has targeted a polymerase to a specific promoter is left behind once transcription begins.

In eukaryotes, the polymerases do not transcribe naked DNA, but instead DNA packaged as chromatin (see the following chapters). This presents the transcribing polymerase with an additional problem not found in prokaryotes and may go some way towards explaining the greater number of subunits in eukaryotic polymerases. However, it is interesting that most of the PIC is left behind when the polymerase starts work, emphasizing that much of its complexity reflects the demands of regulating the initiation step, rather than any additional catalytic functions. Once the polymerase has left the PIC, the way is open for a new polymerase to attach itself to the promoter and reinitiate the process. All the time that TFIID and associated factors remain attached to the promoter, the second and subsequent rounds of transcription will be accomplished much more quickly than the first.

In the previous chapter it was noted that in prokaryotes the transcribing polymerase can sometimes pause, or 'stall', thereby effectively slowing the overall rate of transcription. An extreme example of control through polymerase stalling is also found in eukaryotes. The 'heat shock' genes undergo a rapid up-regulation of transcription within minutes of exposure of the cell to a variety of environmental insults, of which elevated temperature is only one. The best studied is the *HSP70* gene in *Drosophila*. Under normal conditions, the *HSP70* gene contains a PolII complex that has stalled after transcribing about 25 bp of the gene. A temperature increase results in a rapid cascade of events that leads to release of the stalled polymerase and synthesis of complete *HSP70* transcripts. This mechanism is a very efficient way of controlling genes where a rapid up-regulation of transcription is functionally essential.

Experimental considerations

Thus far, little attention has been given to how the information on the composition of eukaryotic PICs has been gathered. In putting in place some general concepts, little attention has been given to the major experimental problems that must be overcome in order to establish these concepts. Some experimental approaches are dealt with in later chapters, but at this early stage a couple of general points are worth making.

The first is that the information presented has come from a combination of biochemical and genetic approaches. Protein complexes such as TFIID have been isolated by conventional biochemical means (such as column chromatography) and assayed by their ability to enhance transcription *in vitro*. In order to characterize the protein components of these complexes, they have been dissociated and the proteins isolated individually. For many of these proteins it has been possible to identify, clone and sequence the genes that encode them. This both provides the exact amino acid sequence of the protein in question and permits the synthesis of relatively large amounts of the 'recombinant' protein in bacteria transformed with the appropriate gene. These recombinant proteins can be used to assemble transcription complexes *in vitro* out of pure components.

The second is that researchers are making extensive use of simple model systems, primarily yeast. Yeast genetics provides an extraordinarily powerful tool for investigating protein function. Its uses range from the relatively simple approach of knocking out (deleting) certain genes and examining the effect on the yeast phenotype, to more complex manipulations in which genes are mutated in specific ways so as to investigate the functional roles of specific protein domains. Examples of these approaches can be found in later chapters. As noted in the previous chapter, the extreme conservation through evolution of many elements of the transcription apparatus means that studies on simple eukaryotes can provide valuable insights into the workings of the analogous systems in higher organisms.

Large genome problems: why are things so complicated?

This is probably a good point at which to try to come up with some sort of justification for the extraordinary complexity of the eukaryotic PIC. Why should something that is achieved with such elegant simplicity (in retrospect) by prokaryotes have attained such monstrous dimensions in even a simple, single-celled eukaryote such as yeast? The answer presumably lies partly in the need to regulate an increased number of genes, a more complex intracellular structure and a more sophisticated range of responses to environmental and nutritional signals. Such responses are likely to require the coordinated expression of larger numbers of genes than can be accomplished through the elegantly simple operon mechanism used by bacteria. Upstream control elements are not unknown in prokaryotes, but they are rare, whereas in eukaryotes they are the norm. In order to do their job, these various activators must be able to enhance assembly of a functional PIC on the gene of which they are a part. It would be wasteful to try to manufacture different PIC complexes to suit the spectrum of activators upstream of different genes (and would also lead us into logical difficulties). It makes more sense to have a universal complex that works (with relatively minor modifications) for tens of thousands of different genes, even though, for any given gene, some of the components may be redundant. A single protein cannot provide this degree of functional flexibility. For example, although TBP itself interacts directly with a number of transcriptional activators, it cannot possibly provide binding sites for all the many different activator and coactivator proteins it is likely to meet, or indeed for the various protein components of the PIC itself.

Co-ordinated gene regulation in eukaryotes

Eukaryotes have available to them a number of gene regulation mechanisms that are not found in prokaryotes. Possession of a nuclear membrane in itself provides an extra level of control by separating the processes of transcription (nuclear) and translation (cytoplasmic). In prokaryotes the two processes occur simultaneously, with translation beginning even before the transcript is complete. In eukaryotes, many mRNAs are degraded before they reach the cytoplasm, allowing fine-tuning of transcript levels. Primary transcripts can also be modified by splicing, a process that can be used to produce several protein products from the information in a single gene. However, of greater importance is the method by which DNA is packaged within the nucleus, i.e. the presence, only in eukaryotes, of the DNA–protein complex called chromatin. In prokaryotes such as *E. coli*, the DNA is largely exposed to proteins present within the cell. Because of this, DNA-binding proteins such as CAP, even if present in only small amounts, will have a measurable effect on the transcription of certain genes. In eukaryotes, this is prevented because: (i) there is at least partial separation of transcription factors, which are made in the cytoplasm, and genes, which are nuclear; and (ii) DNA is usually packaged with specific proteins (as chromatin) in such a way as to inhibit the binding of transcription factors, unless measures are taken to allow this to happen. In addition, further levels of DNA packaging and sequestration can be used to switch off *groups of genes* at particular stages of development or in particular cell types. In other words, DNA packaging as chromatin, along with a complex, multicomponent PIC, provides the sort of co-ordinated expression of groups of genes that, in bacteria, is achieved by putting a small number of genes under the control of the same promoter in an operon. In the following chapter I will introduce the basic elements of chromatin packaging and begin to explore their effects on gene expression.

Further Reading

Evolutionary issues and cell types

Alberts, B., Bray, D., Lewis, J., Raff, M. & Watson, J. D. (1994) *Molecular Biology of the Cell.* 3rd Edition. Garland, New York.

Bird, A. P. (1995) Gene number, noise reduction and biological complexity. *Trends Genet.,* **11:** 94–100.

Sulston, J. E. & Horvitz, H. R. (1977) Postembryonic cell lineages of the nematode worm *Caenorhabditis elegans. Dev. Biol.,* **56:** 110–156.

Genome composition and sequencing

Recently, issues of the journals *Science* and *Nature* have been devoted to descriptions of the genomes of particular organisms. Specific articles are cited, but others in the same issues are also of interest. For the most complete and up-to-date information, consult the listed Web sites:

Yeast (*S. cerevisiae*)

http://genome-www.stanford.edu/Saccharomyces/

Goffeau, A., Barrell, B. G., Bussey, H. *et al.* (1996) Life with 6000 genes. *Science,* **274:** 546–567.

Mewes, H. W., Albermann, K., Bähr, M. *et al.* (1997) Overview of the yeast genome. *Nature,* **387:** 7–9.

Caenorhabditis elegans

http://www.sanger.ac.uk/Projects/C/elegans/

C. elegans sequencing consortium (1998) Genome sequence of the nematode *C. elegans*: a platform for investigating biology. *Science,* **282:** 2012–2018.

Drosophila melanogaster

http://flybase.bio.indiana.edu/

Adams, M. D., Celniker, S. E., Holt, R. A. *et al.* (2000) The genome sequence of *Drosophila melanogaster. Science,* **287:** 2185–2195.

Humans, mice

http://www.ncbi.nlm.nih.gov/genome/guide/

Bouck, J. B., Metzker, M. L. & Gibbs, R. A. (2000) Shotgun sample sequence comparisons between mouse and human genomes. *Nature (Genetics),* **25:** 31–33.

Rubin, G. M. *et al.* (2000) Comparative genomics of the eukaryotes. *Science,* **287:** 2204–2215.

Promoter DNA

Kollmar, R. & Farnham, P. J. (1993) Site-specific initiation of transcription by RNA polymerase II. *Proc. Soc. Exp. Biol. Med.,* **203:** 127–139.

McClure, W. R. (1985) Mechanisms and control of transcription initiation in prokaryotes. *Ann. Rev. Biochem.,* **54:** 171–204.

RNA polymerases and transcription initiation complexes

Nikolov, D. N., Chen, H., Halay, E. D. *et al.* (1995) Crystal structure of a TFIIB–TBP–TATA-element ternary complex. *Nature,* **377:** 119–128.

Roeder, R. G. (1996) The role of general initiation factors in transcription by RNA polymerase II. *Trends Biochem. Sci.,* **21:** 327–335.

Travers, A. (1996) Building an initiation machine. *Curr. Biol.,* **6:** 401–403.

White, R. J. (2000) *Gene Transcription.* Blackwell Science Ltd, Oxford.

Young, R. A. (1991) RNA polymerase II. *Ann. Rev. Biochem.,* **60:** 689–715.

Zawel, L. & Reinberg, D. (1995) Common themes in assembly and function of eukaryotic transcription complexes. *Ann. Rev. Biochem.,* **64:** 533–561.

Chapter 3: The Nucleosome: Chromatin's Structural Unit

Introduction

All organisms, prokaryotic or eukaryotic, must deal with the problem of packaging a relatively long piece of DNA into a small space within the cell. In eukaryotes, DNA is sequestered in a particular subcellular organelle, the nucleus. In the Prologue, the size of the problem was illustrated by scaling things up so that the nucleus was the size of a large grapefruit (ie. about 10 cm in diameter). On this scale, the DNA molecules to be packaged in a typical eukaryotic cell would, in total, be about 20 km long. Admittedly the DNA is thin, even on this scale (about 0.02 mm), and the volume of the container is more than sufficient to accommodate it, but the problem is not a simple packaging one. The overriding requirement is that the DNA, once in place, must be able to function. It must be replicated, with complete accuracy just once (and no more than once) each cell cycle, the two copies must be separated into the two daughter cells each time the cell divides and the genetic information encoded by the DNA must be expressed in a way that is appropriate to the particular cell at the particular stage of development that it has reached.

It seems inevitable that the mechanisms by which DNA is packaged into the nucleus will have a major effect on its function. It is hard to imagine how it could be otherwise. It is also worth remembering that solutions to the structural and functional aspects of the DNA packaging problem must have coevolved so as to accommodate the differing requirements of each. Such coevolution will lead, inevitably, to mechanistic links between the two processes. Moreover, the two major components of DNA function, namely replication and transcription, are also likely to have become increasingly interlinked as a result of adjustments, refinements and compromises during evolution.

Exploring how DNA is packaged in the nucleus

If one isolates nuclei from a sample of cells or a piece of tissue (usually a fairly straightforward procedure) and extracts the proteins, about half, by weight, of all the protein extracted would be made up of members of a single protein family, the histones. Histones have been known about for over 100 years. They were first described, and named, by Albrecht Kossel in 1884. All members of the family are relatively small, most having molecular weights of 10–12 kDa, and highly basic, being rich in lysine and arginine, amino acids that are positively charged at near neutral pH. Over the years since they were first described, histones have been found in all eukaryotic cells in which they have been looked for. They are not found in prokaryotes, although some prokaryotes do have proteins with histone-like structural motifs. As the histones were better characterized, it became possible to distinguish five types within the histone family on the basis of size, net charge, relative content of lysine and arginine and solubility properties. These were at first given different names by different research groups, making the early literature difficult to follow in places, but they are now universally designated H1, H2A, H2B, H3 and H4. Their properties are listed in Table 3.1.

Because they are small and present within cells in relatively large amounts, histones were among the first proteins to be purified and sequenced. As sequence data on histones from different species have accumulated, it has become clear that the primary structure (i.e. amino acid sequence) of any given histone is often virtually identical from one organism to another. Histones H3 and H4 are among the most highly conserved proteins known.

The weight of histones in the cell nucleus is approximately equal to the weight of the other major nuclear component, DNA. Until about 50 years ago, the roles of these two nuclear components were a matter of debate, with those who felt that DNA was far too simple a polymer to encode genetic information turning to the histones as more likely repositories. The elucidation of the double helical structure of DNA, the cracking of the genetic code and the demonstration of its near universality rapidly cleared away any lingering doubts about the role of DNA, but left the histones at something of a loose end. It was easy to talk in vague terms about a 'structural' role and their positive charge did indeed make them excellent DNA-binding proteins. However, it was unclear how this putative role might be accomplished, particularly as the mixing of histones and DNA in a test tube under physiological ionic conditions resulted not in a progressive organization and compaction of the DNA polymer, but in the almost instantaneous appearance of an unstructured, insoluble precipitate.

The eventual unravelling of the way in which histones and DNA are complexed in the cell nucleus came about as a result of observations from a number of workers using different experimental approaches. It is a story that has

Table 3.1 Some properties of histones.

Histone	Number of residues	Molecular mass	Residues/mol (%)		Net charge
			Lysine	Arginine	
Core					
H2A	129	13960	14 (10.9)	12 (9.3)	+15
H2B	125	13774	20 (16.0)	8 (6.4)	+19
H3	135	15273	13 (9.6)	18 (13.3)	+20
H4	102	11236	11 (10.8)	14 (13.7)	+16
Linker					
H1	224	22500	66 (29.5)	3 (1.3)	+58

received less popular attention than that surrounding the discovery of the DNA double helix, but that is every bit as exciting and, arguably, equally important. I shall not make any attempt to describe the twists and turns of the story here and will focus only on those key developments that led directly to our present model of chromatin structure. This inevitably involves glossing over complications and runs the risk of making the whole process seem much simpler than it actually was. A fuller description from someone who was closely involved will be found in the book by Kensal Van Holde (1988, see Further Reading).

The subunit structure of chromatin

The elucidation of the subunit structure of chromatin resulted from the coming together of results from different individuals and research groups, sometimes working together and sometimes completely unaware of developments elsewhere. One approach that proved crucial in the early years, and has remained central to chromatin research ever since, is controlled digestion by endonucleases, i.e. enzymes that make cuts within the DNA polymer (in contrast to exonucleases that nibble away at the ends). A short paper in 1973 showed that rat liver nuclei contained an endogenous, endonuclease activity that, when intact nuclei were incubated under appropriate conditions, cut DNA at regular, repeating intervals. The authors suggested that '. . . chromatin has some simple, basic, repeating substructure with a repetitive spacing of sites that are potentially accessible to . . . endonuclease'. An interpretation of these results (benefiting from hindsight, but not too far from what was originally proposed) is shown in Fig. 3.1. It was soon shown that the same pattern of DNA

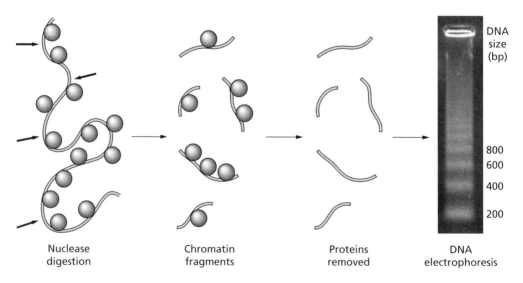

Nuclease   Chromatin   Proteins   DNA
digestion   fragments   removed   electrophoresis

Fig. 3.1 DNA fragments generated by exposure of chromatin to nucleases have sizes that are multiples of about 200 bp. Thick arrows illustrate possible nuclease cutting sites.

cleavage could be achieved in nuclei that lacked the endogenous nuclease by incubating them with the bacterial enzyme, micrococcal nuclease, under conditions that allowed it to enter the nucleus. Most nuclear DNA (85% in the original work) was susceptible to this very specific pattern of cleavage, showing that the repeat packaging unit, whatever it might be, served to package most of the nuclear DNA rather than just a minor subfraction. This conclusion did not come as a complete surprise. It was consistent with earlier data from analysis of whole nuclei by X-ray diffraction (the same technique that was used to define the double-helical structure of DNA). This work had shown that there were regularities in the diffraction patterns of whole nuclei and chromatin preparations that could not be explained by the DNA or proteins alone, and that suggested a regular, repeating structure of some sort. The enormous complexity of whole nuclei or chromatin preparations compared with the crystals normally used for X-ray diffraction studies precluded more detailed interpretation of the data.

Having established the probability that chromatin consisted of a basic, repeating subunit, the next step was to find ways of isolating this subunit in more or less intact form so that it could be subjected to structural and biochemical characterization. This necessitated treating nuclei under conditions that were sufficiently disruptive to extract chromatin with its subunit structure intact, but not so harsh as to dissociate it completely. One such approach was developed by electron microscopists attempting to visualize chromatin structures directly. Looking at whole nuclei or sections of nuclei gave little useful information and a variety of (more or less mild) disruptive techniques were tried. The major problem was, as always, the danger that the disruption and/or fixation procedures were themselves distorting the original structure. The breakthrough came when a procedure originally developed for visualizing transcription complexes in the giant, lampbrush chromosomes of amphibian oocytes was applied to the study of chromatin. Nuclei (from various tissues) were isolated under isotonic conditions (i.e. 200 mM KCl) and then exposed to reduced salt concentration to induce swelling and gradual lysis. Nuclei lysed under these relatively mild conditions were fixed with formaldehyde and spread onto carbon-coated grids for EM analysis. Strands of material could be seen spilling out of the lysed nuclei. In a seminal paper, Donald and Ada Olins described the 'beads-on-a-string' appearance of these fibres, in which particles of 60–80 Å diameter were separated by a 15-Å fibre (Fig. 3.2). Experiments carried out by others in parallel, or slightly later, and using slightly different spreading, fixation and visualization techniques, gave essentially the same picture, although the particles were slightly larger, 100–130 Å. The particles were originally called v-bodies by the Olins and were proposed to consist of aggregates of the five histones and DNA, linked by naked DNA (the 15 Å fibre visible in favourable spreads). The simplicity of these structures contrasted with the variety of fibre structures revealed by other procedures for preparing chromatin for EM analysis. The result also provided a very neat interpretation of the nuclease digestion results, with

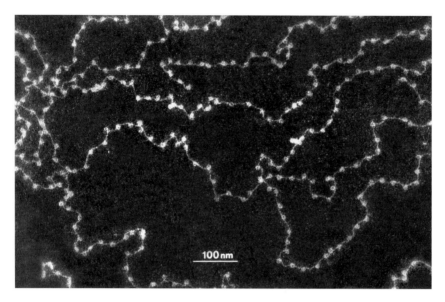

Fig. 3.2 Darkfield electron micrograph showing nucleosomes and linker DNA. Chicken erythrocyte nuclei were lysed in dilute salt solution. Released chromatin was fixed in formaldehyde, spread onto a carbon-coated grid by centrifugation and stained with uranyl acetate. (Photograph provided by Donald and Ada Olins.)

the proposition that the DNA within the particles was resistant to nuclease cutting while that in between was sensitive. The size of the particles also corresponded remarkably well with the fact that the major repeat detected by X-ray diffraction of whole nuclei and chromatin was at 100 Å.

It is always nice to have a *picture* of structures proposed on the basis of biochemical tests, but microscopical analysis has two inevitable drawbacks. The first is that the structures observed may be artefacts of the preparation, fixation and spreading procedures used and the second is that electron microscopy does not, in itself, provide any data on the composition of the structures observed. This requires their isolation and biochemical characterization. Once again, nuclease digestion proved particularly useful in addressing these problems. If nuclei are incubated with micrococcal nuclease and the chromatin is then isolated under mild conditions, it can be separated by centrifugation on sucrose density gradients into a series of fractions of increasing size (Fig. 3.3a,b). The peaks on the gradient are particularly clear if histone H1 is dissociated from the chromatin by exposure to moderate salt concentrations (around 0.4 M). The peak fractions isolated from sucrose gradients contain equimolar amounts of H2A, H2B, H3 and H4, and the size of the DNA in each fraction corresponds exactly to that of the individual bands in the DNA ladder generated by digestion of nuclei with either endogenous or added nucleases (Fig. 3.3c,d). These experiments not only define the composition of the chromatin subunit, but also tell us at once that histone H1 is not an essential structural component of this subunit.

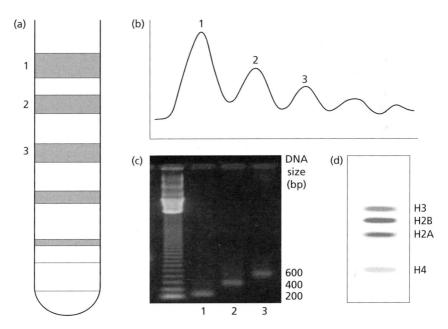

Fig. 3.3 Chromatin fragments can be separated into units of discrete size by centrifugation. (a) If a chromatin preparation (made as outlined in Fig. 3.1) is layered on the top of a sucrose solution in a centrifuge tube and then spun at about 100 000 g, chromatin fragments will move down the tube at different speeds, with the larger ones moving fastest. After a few hours, discrete bands of chromatin will have separated in the sucrose solution. (b) The chromatin bands are detected by separating the solution in the tube into small fractions (taken from the top or the bottom) and measuring their optical density. (c) If the DNA from each peak fraction is resolved by electrophoresis, it will form a single band of a discrete size (ethidium bromide stain). (d) Protein electrophoresis shows that each fraction contains the same set of four core histones in equimolar amounts (Coomassie Blue stain; H4 binds this dye less well than the other three histones and so stains less intensely).

In 1974, Roger Kornberg brought together the disparate bits of information outlined above, together with some additional data on histone–histone interactions *in vitro*, and proposed an elegantly simple model for the structure of chromatin. This model has provided the basis for our ideas on chromatin structure and function ever since. He suggested that:

1 chromatin consists of a fundamental repeating unit made up of 200 base pairs of DNA and *two each* of the four histones H2A, H2B, H3 and H4; i.e. the histones formed an eight-subunit (octameric) structure. The stoichiometry was initially based on the relative amounts of the different histones in various species, and on histone–histone cross-linking experiments. It was confirmed by analysis of isolated chromatin particles by Pierre Chambon and co-workers, who named the particle the nucleosome. This name has been universally adopted.

2 A chromatin fibre consists of many such nucleosomes forming a flexibly jointed chain. This corresponds to the beads-on-a-string fibres observed under the EM.

The structure of the nucleosome

Analysis of histone–histone interactions within the nucleosome by chemical cross-linking showed that the eight histones were all connected in a single particle (the histone octamer) that could be dissociated, by manipulation of ionic conditions, into an $(H3–H4)_2$ tetramer and two H2A–H2B dimers. However, it was not immediately obvious exactly how the DNA was organized in relation to the histones.

The first experimental clues came, once again, from nuclease digestion studies when a careful analysis of how nucleosomal DNA was digested by the endonuclease DNaseI led to the proposal that the DNA was coiled around the *outside* of the histone octamer. At first sight this seems at odds with the fact that the histones *protect* the DNA from digestion with micrococcal nuclease or indeed endogenous nucleases. The explanation lies in the mechanisms by which different endonucleases cleave DNA. Micrococcal nuclease makes double-stranded cuts in DNA and, like bolt-cutters, the enzyme needs to straddle the DNA before it can cut it (Fig. 3.4). Because of this, the enzyme is unable to cut DNA resting on a surface. This explains why nucleosomal DNA is protected from cutting by micrococcal nuclease, even though it is located on the outside of the histone octamer (Fig. 3.4). DNaseI operates rather differently: where it attaches to the DNA it makes cuts (nicks) in one or the other of the two strands of the DNA double helix. When the DNA is laid along a surface (either a perfectly flat crystal surface in the lab or a curved, protein surface, as in the nucleosome), DNaseI can still make these single-strand nicks, but only the DNA furthest from the surface can be cut easily. This is shown diagrammatically in Fig. 3.5. This gives rise to a cutting pattern in which single-strand cuts appear most frequently at intervals of about 10 bases, i.e. about once per helical turn (Fig. 3.5). This DNaseI cutting pattern is characteristic of nucleosomal DNA and is frequently used as a diagnostic test for the presence of nucleosomes over specific regions of DNA. The conclusions drawn from the nuclease digestion studies were confirmed by the completely independent technique of neutron scattering. This procedure can distinguish signals generated by DNA and protein, and studies of nucleosomes in solution confirmed that nucleosomal DNA is indeed wrapped around the outside of a histone core.

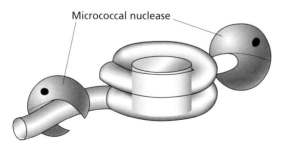

Micrococcal nuclease

Fig. 3.4 Micrococcal nuclease must surround DNA in order to cut it, so DNA in the nucleosome core is protected, even though it lies on the external surface.

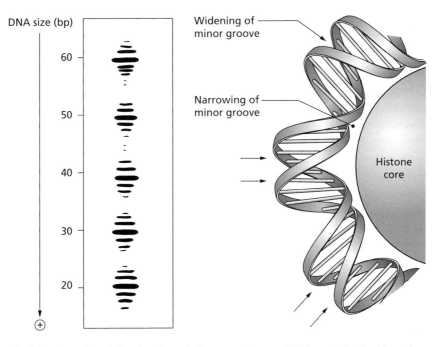

Fig. 3.5 The nuclease DNase1 preferentially cuts nucleosomal DNA at sites furthest from the histone core. Because of preferential, single-strand cutting at the sites indicated by arrows, the most frequent pieces of DNA generated by DNaseI treatment of nucleosomes will differ in size by about 10 bp (i.e. one turn of the DNA helix). This periodicity of cutting can be detected by DNA electrophoresis.

Thus, by the mid-1970s, the composition and basic structure of the fundamental subunit of chromatin structure was established. It was also becoming clear that the basic structure of the nucleosome was essentially the same in even the most distantly related eukaryotes, from single-celled microorganisms such as yeasts to higher plants and to mammals. However, a definitive description of the nucleosome came only when it could be subjected to direct structural analysis. This was made possible by X-ray crystallography. The technical difficulty that stood in the way of using this powerful approach to its full potential was the need to prepare high-quality nucleosome crystals. The primary distinguishing feature of crystals is their regularity of structure, and it is virtually impossible to generate satisfactory crystals from starting material that is a heterogeneous mixture. Unfortunately, nucleosomes, as originally defined, fall into this category. Even mono-nucleosomes purified on sucrose gradients are still heterogeneous, mainly, but not exclusively, because of the slightly differing lengths of linker DNA associated with each one. This is a consequence of the inevitable variation in exactly where the nuclease happens to cut (see Fig. 3.1). To overcome this, nuclease digestion was carried out until only that DNA directly associated with the histone octamer (146 bp in all) remained uncut, i.e. all the linker DNA had been removed but the cutting had not gone so far as to begin to degrade the particle itself. This structure, the nucleosome core particle, was eventually crystallized by Klug and coworkers at the MRC Laboratory of Molecular Biology, Cambridge, and provided the basic raw material for more detailed structural analysis. X-ray crystallography, after several years of painstaking experimentation, gave a structure at a resolution of 25 Å, which was published in 1977 and which gave the first, definitive picture of the fundamental structure of the nucleosome core particle. A refined structure, at 7 Å resolution, was published in 1984. The experiments revealed a histone core around which are wrapped 146 base pairs of DNA in $1\frac{3}{4}$ superhelical turns. The core is shaped roughly like a flattened cylinder, 110 Å in diameter and 55 Å high. It is slightly wedge shaped, causing the stacks of core particles to curve in the crystals. There is an axis of two-fold symmetry, the dyad axis. The structure is shown diagrammatically in Fig. 3.6(a).

Despite the triumphs of X-ray crystallographic analysis, fundamental questions remained. One such concerned the role of histone H1. The fact that most studies of nucleosome structure employed chromatin from which H1 had been removed by salt extraction showed that it was not an essential structural component. However, careful analysis of the kinetics of chromatin digestion by micrococcal nuclease indicated that H1 was closely associated with the nucleosome *in vivo*. When native chromatin, complete with H1, was digested, the smallest chromatin fragment initially contained 165 base pairs of DNA. Only after further digestion was this reduced to 146 base pairs. The particle containing 165 bp of DNA could be isolated and shown to contain H1, whereas the 146 bp particle did not. It was concluded that H1 protected, albeit transiently, about 20 bp of nucleosomal DNA from nuclease digestion. The linker histone, H1, is thought to occupy a position outside the core octamer, close to where the DNA strand enters and exits the nucleosome. This is consistent with its effects on nuclease digestion and with the observed stoichiometry of one H1 molecule per nucleosome. A nucleosome complexed with H1, sometimes referred to as a

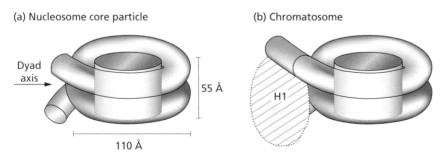

(a) Nucleosome core particle   (b) Chromatosome

Dyad axis

55 Å

110 Å

H1

Fig. 3.6 Dimensions of the nucleosome core particle (a) and possible location of linker histone H1 in the chromatosome (b).

'chromatosome', is shown in Fig. 3.6(b). H1 has a globular core with two relatively long N- and C-terminal tails capable of interacting with both linker and
core DNA. The exact structural relationship between H1 and the nucleosome is
still a matter for debate, and will remain so until it is possible to crystallize
chromatosomes.

The high-resolution structure of the core particle

In 1997, the structure of nucleosome core particle crystals at 2.8 Å resolution
was published, giving a remarkable increase in structural detail and providing
valuable insights into the organization of both the histone and DNA components. Structural motifs in the core histones are readily identified and the path
and structure of the core DNA are defined in atomic detail. The structure is
shown in Plates 3.1 and 3.2. Representation of complex structures at this level
of detail has been greatly facilitated by advances in computer graphic techniques, but the main reason for the improvement has been the use of a homogeneous population of core particles for crystallization. This was achieved by
the *in vitro* assembly of core particles from a recombinant DNA fragment of uniform length and recombinant histones (i.e. histones expressed in bacteria from
inserted plasmids containing the amphibian genes for H2A, H2B, H3 or H4). Recombinant histones do not contain the many post-translational modifications
present in native histones and, because of this, provide more uniform core particles and crystals. The long periods of time between the successive refinements
of the crystal structure give some indication of the amount of experimental
effort involved in this sort of analysis.

Histone–histone interactions in the nucleosome

Some of the most important pieces of experimental data used in putting
together the first model of the nucleosome concerned the ways in which histones could associate with one another *in vitro*. Certain histones readily associated to form heterodimers, namely H3–H4 and H2A–H2B. The H3–H4 dimer
could further associate to form a tetramer $(H3–H4)_2$. An early clue that these associations were relevant *in vivo* came from the observation that careful dissociation of chromatin fragments by progressively increasing the salt concentration
led to the loss of H2A–H2B dimers, leaving the DNA constrained only by
$(H3–H4)_2$ tetramers.

Much more detail about these interactions was provided by X-ray crystallography of core octamers in the absence of DNA, primarily from the lab of E.
Moudrianakis. These studies defined elements of secondary structure, primarily α-helices, within the histone octamer, and identified two crucially important structural motifs—the histone fold and the histone handshake. They are
illustrated in Fig. 3.7. The histone fold region comprises a sequence of three α-

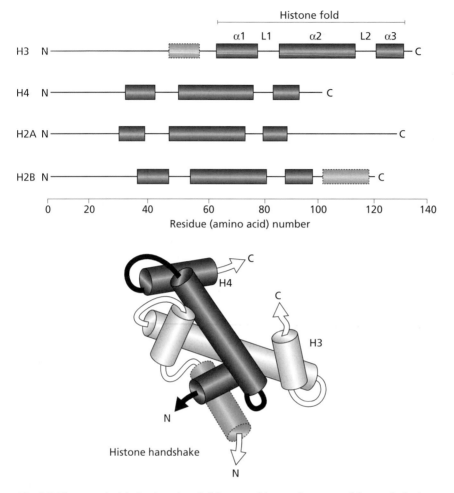

Fig. 3.7 The central, globular domains of all four core histones have a set of three α-helical regions that form a characteristic structure known as the histone fold. The shaded boxes labelled α1, α2 and α3 in the upper part of the figure represent the three α-helical regions of the histone fold domain. L1 and L2 are nonhelical loops between the helices. The broken lines represent α-helical regions not forming part of the histone fold. The fold domains of H3 and H4 interact in the core particle to form the 'histone handshake', shown in the lower part of the figure. The fold domains of H2A and H2B interact in the same way.

helices, one long and two short, with nonhelical 'loop' regions separating them. This characteristic arrangement of α-helices is found in all four-core histones, and in some nonhistone proteins as well. In the core octamer, the three helical regions in each of the four histones are arranged in a very similar way to give the tertiary structure known as the histone fold (Fig. 3.7). The histone folds of H3 and H4 and of H2A and H2B interlock to give a structure described as the 'histone handshake' (Fig. 3.7). It is this handshake interaction that gives rise to the stable H3–H4 and H2A–H2B dimers that are fundamental in core octamer assembly. Both the histone folds and histone handshakes are readily seen in great detail in high-resolution models of the nucleosome core particle (Plate 3.2).

Nucleosomal DNA

The path followed by the DNA double helix across the surface of the histone octamer has been defined by a combination of high-resolution X-ray crystallography, nuclease digestion and histone–DNA cross-linking. The DNA makes almost two turns around the histone core forming a *left-handed* superhelix. The double helix of the DNA itself is *right-handed* (the difference is illustrated in Fig. 3.8). The path and structure of the DNA around the histone octamer is clearly resolved in the 2.8 Å structure. The DNA does not follow a smooth, continuous curve, but is bent more sharply in some regions than in others. This bending inevitably distorts the double helix, with the extent of this distortion varying with the degree of bending. The uneven bending of core DNA can be seen in Plate 3.2, which shows the path of just one turn of DNA around the histone core. Distortion of the double helix will, potentially, influence some of the functional properties of the nucleosomal DNA, such as the ability of sequence-specific DNA-binding proteins to recognize and attach to their cognate sequences or of nucleases to cut. As will be discussed in later chapters, the path of the DNA can be an important element in the role of the nucleosome as a regulator of transcription. The relative resistances of different DNA sequences to bending can be

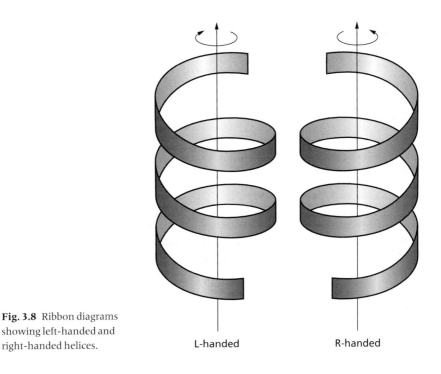

Fig. 3.8 Ribbon diagrams showing left-handed and right-handed helices.

L-handed   R-handed

a major factor in determining exactly where nucleosomes are positioned along DNA.

One property of core particle DNA that has caused some controversy and concern is its twist, i.e. the number of base pairs per turn of the helix. In B-form DNA (the usual form found in solution under isotonic conditions), there are 10.6 base pairs per turn. In the nucleosome, it seems likely that this is reduced, overall, to about 10 bp per turn, i.e. the DNA is overwound (Fig. 3.9). This is not an easy property to measure, requiring particularly careful analysis of DNaseI digestion patterns of the type shown in Fig. 3.5. Analysis of the 2.8 Å crystal structure gives an overall twist of 10.2 bp per turn, but shows very clearly that the twist can vary from place to place on the nucleosome surface, i.e. twist can be a very *local* property. Such differences are apparent in the image shown in Fig. 3.9. It is not hard to appreciate how local changes in the twist of the DNA might influence the ability of nucleases or DNA-binding proteins to access their cognate sequences. It is important to note that the *exact* path of the DNA around the core octamer and its *local* properties may vary depending on the properties of the core histones and the sequence of the DNA. They need not be immutable from one nucleosome to the next.

Histone–DNA interactions

Early attempts to define the points at which the DNA superhelix interacts with the histone octamer involved the use of DNA–histone cross-linking techniques and served as an invaluable guide in assembling the first, low-resolution structure for the nucleosome core particle. The latest X-ray crystallographic structures allow interactions between specific amino acids and the DNA backbone to be identified. The histone fold domains are responsible for binding most of the DNA (121 bp). Each of the four histone heterodimers that make up the core particle (i.e. two H2A–H2B and two H3–H4) binds about 30 bp of DNA. Binding is primarily between amino acids in the α-helices or intervening loop regions and the phosphodiester backbone of the DNA, something that allows the core octamer to bind with equal efficiency to the almost infinite variety of DNA

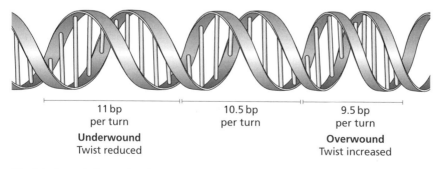

| 11 bp per turn | 10.5 bp per turn | 9.5 bp per turn |
| **Underwound** Twist reduced | | **Overwound** Twist increased |

Fig. 3.9 Overwinding and underwinding of the DNA double helix.

sequences that it will encounter *in vivo*. Contacts occur every time the minor groove faces the histone core and can involve hydrogen bonds and salt links to DNA phosphates and nonpolar contacts with deoxyribose groups. In addition, in each of the 14 instances where the minor groove contacts the core, it is penetrated by the side-chain of an arginine residue located in either a histone fold domain (10 times) or an N-terminal tail region (4 times, see Chapter 4).

Nucleosome assembly and disassembly

The combination of histone–DNA contacts that leads to assembly of the nucleosome core particle give it a generally consistent and stable structure. However, like many macromolecular complexes, the nucleosome is sensitive to even small shifts in ionic strength, with partial unfolding detectable at ionic strengths below 1 mM and above 300 mM. Low ionic strength, or exposure to reagents such as urea, will dissociate the histone–histone interactions (some of which involve hydrophobic bonds) that stabilize the core octamer. High ionic strengths, salt concentrations of 800 mM and above, will break the primarily ionic bonds between DNA and histones and lead to complete dissociation of the core DNA. However, even at salt concentrations between 100 and 300 mM (concentrations that bridge the isotonic), there is detectable, local (i.e. not total) dissociation of DNA from the histone octamer. This is probably mainly the result of DNA peeling away, and subsequently rejoining, at the points where it enters and leaves the core particle. This sort of dynamic equilibrium is important when considering the way in which proteins such as transcription factors bind to DNA packaged as nucleosomes (Chapter 5). As one might expect, H1 has a strong stabilizing effect, serving to keep the DNA ends attached to the octamer (Fig. 3.6b).

Despite all this, the nucleosome core particle is a structure that can form spontaneously. It has the ability to self-assemble, under the right conditions, and must therefore be energetically favourable. Isolated core particles can be completely dissociated by a combination of 2 M salt, which will break the DNA–histone bonds, and 5 M urea, which will dissociate the histone octamer and denature the component histones. If the solution is then dialysed to bring the mixture gradually back to low salt conditions, core particles will re-form. These particles are indistinguishable from the starting material, at least in terms of general measures of structural integrity such as sedimentation coefficient, nuclease digestion and appearance under the EM. The ability of the core particle to reassemble from dissociated components is frequently taken advantage of in experimental protocols designed to build core particles with defined histone and DNA content.

Further Reading

General

Van Holde, K. E. (1988) *Chromatin*. Springer, New York.
Wolffe, A. (1998) Chromatin: Structure and Function. 3rd Edn. Academic Press.

Early days

Hewish, D. R. & Burgoyne, L. A. (1973) Chromatin substructure. The digestion of chromatin DNA at regularly spaced sites by a nuclear deoxyribonuclease. *Biochem. Biophys. Res. Comm.*, **52:** 504–510.
Kornberg, R. D. (1974) Chromatin structure: a repeating unit of histones and DNA. *Science*, **184:** 868–871.
Kornberg, R. D. & Klug, A. (1981) The nucleosome. *Sci. Am.*, Feb 1981: 52–64.
Noll, M. (1974) Subunit structure of chromatin. *Nature*, **251:** 249–251.
Olins, A. L. & Olins, D. E. (1973) Spheroid chromatin units (ν bodies). *Science*, **183:** 330–332.

High-resolution nucleosome structure

Arents, G., Burlingame, R. W., Wang, B.-C. *et al.* (1991) The nucleosome core histone octamer at 3.1 Å resolution: a tripartite protein assembly and a left-handed superhelix. *Proc. Natl. Acad. Sci., USA*, **88:** 10148–10152.
Finch, J. T., Lutter, L. C., Rhodes, D. *et al.* (1977) Structure of the nucleosome core particle of chromatin. *Nature*, **269:** 29–36.
Luger, K., Mäder, A. W., Richmond, R. K., Sargent, D. F. & Richmond, T. J. (1997) Crystal structure of the nucleosome core particle at 2.8 Å resolution. *Nature*, **389:** 251–260.
Pruss, D., Hayes, J. J. & Wolffe, A. P. (1995) Nucleosome anatomy—where are the histones? *BioEssays*, **17:** 161–170.
Rhodes, D. (1997) The nucleosome core all wrapped up. *Nature*, **389:** 231–233.
Richmond, T. J., Finch, J. T., Rushton, B., Rhodes, D. & Klug, A. (1984) Structure of the nucleosome core particle at 7 Å resolution. *Nature*, **311:** 532–537.

Chapter 4: Histone Tails: Modifications and Epigenetic Information

Introduction

In the previous chapter, I discussed the nucleosome core particle primarily in the context of its role in DNA packaging. Emphasis was given to the extreme conservation through evolution both of the structure itself and of the histones from which it is built, and how DNA is configured through specific interactions between the DNA backbone and specific amino acids in the histone core. However, important though this is, it is not the only role of the nucleosome. The particle has a second function that gives it an importance far beyond its initial packaging role and that will appear repeatedly in the following chapters. This is its ability to carry *epigenetic* information.

By epigenetic information, I mean instructions as to how the *genetic* information encoded by the DNA base sequence is to be used. For example, a typical epigenetic instruction might be 'do not transcribe this gene' or 'transcribe this gene only if . . .'. Trying to understand how these instructions are put in place, modified and transmitted from one cell generation to the next are issues that underlie most of the material in this book, and the nucleosome core particle is the first place to look for answers. At first sight, it does not seem a very promising repository for information of any sort, given the evolutionary conservation both of the particle itself and of the histones from which it is assembled. Where does it get the variability needed in order to issue instructions about gene expression? The answer lies in specific regions of the core histones that are not seen in X-ray structures of core particle crystals, namely the regions of up to 25 or so amino acids at the amino-terminal (N-terminal) ends of each of the eight core histones. The tails are exposed on the surface of the nucleosome and do not adopt fixed structures in core particle crystals. As with the core histones in general, the amino acid sequences of the tail domains have been highly conserved through evolution. However, the histone tails, unlike the globular, core domains, are subject to a wide variety of enzyme-catalysed, post-translational modifications. Specific amino acid side-chains can be modified by attachment of specific chemical groups (e.g. phosphate or acetate), changing both their charge and conformation. So, despite their evolutionary conservation, the tail domains show enormous variability from one core particle to another in the spectrum of post-translational modifications that they carry. These modifications constitute a major source of epigenetic information.

In the rest of this chapter, I deal with the structures of the histone tails and outline the various post-translational modifications to which they are subjected. The enzymes that put these modifications in place are dealt with in more

detail in Chapter 8 and the ways in which the modifications can influence transcription and other chromatin functions will crop up repeatedly in subsequent
chapters.

The histone tails

The N-terminal domains are not fully detected by X-ray crystallography, presumably because they do not take up a single, fixed structure in core particle
crystals. They either could remain relatively mobile, or could adopt a variety of
different structures with no single structure predominating. The tail domains
are all rich in positively charged amino acids (Fig. 4.1) and their amino acid
compositions overall do not favour the adoption of any of the usual secondary
protein structural motifs. For example, the high frequency of glycines in H4 and
H2A and of prolines in H2B does not favour extensive α-helix formation. The
tail domains are usually said to have 'random coil' structures. However, this is
almost certainly an oversimplification. Short regions of the tails are quite capable (on the basis of sequence analysis) of adopting an α-helical structure and
evidence has been presented to suggest that, under some circumstances, they
may in fact do so (Fig. 4.1). The close proximity of the tails to core DNA may help

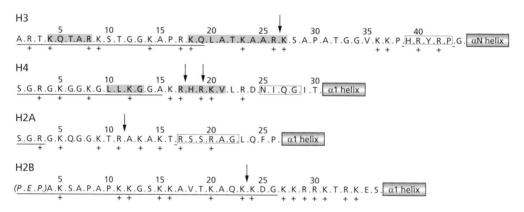

Fig. 4.1 Amino acid sequences (single letter code) of the N-terminal tail domains of the four
core histones. Residue numbers are given above each sequence and positively charged
residues (lysines, K and arginines, R) are indicated below. The vertical arrows indicate the
sites furthest from the N-terminus at which the protease trypsin can cut the tails in
chromatin. These sites define the extent of the exposed N-terminal domains. Sites located
further towards the C-terminal end are protected from cutting by core DNA. Underlined
regions have not yet been located by X-ray analysis of nucleosome core particle crystals and
may be unstructured, or capable of adopting multiple structures. However, on the basis of
their amino acid sequence, some regions in H3 and H4 have a high probability of adopting an
α-helical structure (shaded). Boxed regions in H2A and H4 are short stretches of known α-
helix and the boxed region in H3 is the stretch of amino acids that traverses the core DNA. The
P.E.P. residues at the N-terminus of H2B are missing from the recombinant H2B used to make
core particle crystals (Plate 2.1).

the formation of secondary structures, even where the amino acid composition is not favourable.

The high-resolution picture of the nucleosome has revealed some details of those regions of the N-terminal tail domains that lie closest to the histone folds and has located the positions at which some of the tails leave the core particle. Remarkably, the H3 and H2B tails exit between the two DNA strands through narrow channels formed by the precise alignment of the minor grooves. H2A and H4 tails exit through juxtaposed minor grooves on the top or bottom of the particle. The end result is that an N-terminal histone tail emerges about once every 20 base pairs (Fig. 4.2).

A number of years ago it was shown that controlled digestion of nucleosome core particles with proteases such as trypsin destroyed the N-terminal domains from all core histones, and also cleaved a smaller number of C-terminal residues from H2A (11 or fewer depending on the species) and H3. The remainder of each histone was left uncut. The tail domains were also shown to be accessible to antibodies directed against N-terminal regions and against the C-terminal end of H3. It was concluded that the protease-sensitive regions are exposed on the nucleosome surface rather than buried in the DNA, a conclusion that is

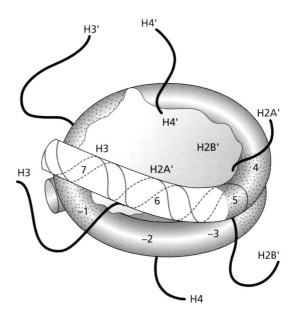

Fig. 4.2 Nucleosome core particle and the histone tail domains. The diagram is loosely based on the X-ray structure of core particle crystals. Complete turns of the DNA helix are numbered from the dyad axis, 1–7 in one direction and −1 to −7 in the other. Both N-terminal tails of H3 and H4 are shown, and just one each for H2A and H2B. The sites at which the tails emerge from the core DNA are based on crystallographic analysis. The length of each tail is defined by the position of the most distal trypsin cutting site (Fig. 4.1) and assumes that each tail is fully extended at 3.7 Å per amino acid. In fact, the tails are likely to adopt defined structure *in vivo*, if only through interactions with other proteins or with core DNA. Stippled regions of core DNA become more sensitive to nucleases after removal of histone tails.

entirely consistent with the high-resolution crystal structure. Despite the fact that the N-terminal regions comprise a significant proportion of the histone content of the core particle (28%), their removal by proteolysis causes no major changes in its physical properties, suggesting that its basic structure is essentially unchanged. Furthermore, if the histone tails are removed and the nucleosome is then dissociated into free DNA and histones by exposure to high concentrations of salt and urea, apparently normal nucleosomes will re-form when the salt and urea are removed by careful dialysis. So the histone tails are not necessary for either maintaining nucleosome structure or assembling nucleosomes, at least *in vitro*. This apparent lack of a structural role contrasts with the fact that the tail domains, particularly of H3 and H4, have been extremely conserved through evolution, suggesting that they must have an important role. So what is it?

One clue came from experiments on the ability of chromatin to form higher-order structures *in vitro*. As discussed in the next chapter, the nucleosome, despite its fundamental importance, does not get us very far in terms of compacting enormously long pieces of DNA into the cell nucleus. Higher levels of folding are required and numerous attempts have been made to model these *in vitro*. It was noted that lengths of chromatin could be induced to switch from the 100 nm, beads-on-a-string conformation, to a higher level of folding (visualized by electron microscopy) only if the tail domains were intact. Their removal by proteolysis effectively suppressed formation of higher-order structures. Once again, this finding is consistent with a feature of the high-resolution crystal structure, namely that one of the two H4 N-terminal tails (residues 16–24) makes contact with a negatively charged patch formed by H2A and H2B residues on an adjacent core particle. Although there is still the possibility that the contact is a crystallization artefact of no significance *in vivo*, it is worth considering the possibility that such internucleosomal links may facilitate the assembly of higher-order structures.

Histone modifications

A second clue about the possible role of the tails comes from the fact that many of the amino acids within the N-terminal histone tails can be modified by specific enzyme activities *in vivo*. This became clear as a result of some of the very first sequencing studies, which showed that particular lysine residues in the tail domains of H3 and H4 were often modified by attachment of an acetate group. Subsequent studies have shown that the tail domains of the core histones can also be modified by phosphorylation, methylation and ADP-ribosylation. The shorter, exposed C-terminal domains of H3 and H2A are not modified, with one exception—the attachment of a small peptide, ubiquitin, to lysine residue 119 of H2A. Interestingly, this residue falls just within the trypsin-sensitive (i.e. exposed) C-terminal region of H2A.

Each of these modifications is carried out, and reversed, by specific enzymes or families of enzymes. These modifications, like the histones themselves, have been conserved through evolution. For example, all species that have been tested so far have the ability to acetylate histone H4. It is worth noting that from an experimental point of view, these modifications make a significant contribution to the heterogeneity of nucleosomes. Even if one were to isolate core particles all with exactly the same length of DNA attached, the enormous variety of histone modifications would still leave one with a mixed population. Only the use of recombinant histones made by genetically engineered bacteria has finally overcome this problem.

Clearly cells are investing a lot of energy into modifying their histones, both in producing the enzymatic machinery to do the job and in providing the metabolic energy needed for each modification step. It is reasonable to conclude that the modifications must play significant roles in regulating chromatin function, and evidence is now accumulating to show that this is indeed the case. What follows is intended only as a brief introduction to each of the post-translational histone modifications. The subject, particularly with regard to histone acetylation, will crop up regularly in the following chapters and details will be added as the story develops.

Acetylation

Acetylation of the histones that organize the nucleosome core particle is a ubiquitous post-translational modification found in all animal and plant species so far examined. Acetylation occurs at the ε-amino groups of specific lysine residues, all of which are found in the amino-terminal domains of the core histones (Fig. 4.1, Table 4.1). It is important at this stage to distinguish between the acetylation of lysine ε-amino groups and the acetylation of the α-amino group of N-terminal serine residue that occurs during the synthesis of H2A, H4 and many other proteins. The latter does not change once it has occurred and, while it may be important for synthesis, processing or transport of the proteins involved, it does not, as far as we know, have any role to play in regulation of chromatin structure.

The transfer of acetate groups from acetyl-CoA to histones and their subsequent removal is catalysed by specific enzymes, the histone acetyl-transferases and deacetylases (Fig. 4.3). The level of acetylation of histones at particular genomic sites is the net result of the equilibrium between these two types of

	Histone	Lysines that can be acetylated
Table 4.1 Acetylation sites in mammalian histones.	H2A	5, 9 (minor)
	H2B	5, 12, 15, 20
	H3	9, 14, 18, 23
	H4	5, 8, 12, 16

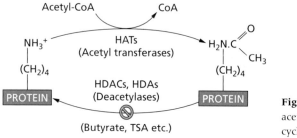

Fig. 4.3 The histone acetylation–deacetylation cycle.

enzyme. Both enzymes tend to be complex, multisubunit structures, whose composition and properties are only now being unravelled. Some of the subunits have been found to be encoded by genes previously identified, usually through genetic studies, as important regulators of gene expression. These important enzymes are described in more detail in Chapter 8.

The steady-state level of acetylation and the rates at which acetate groups are turned over vary between different cell types, with half-lives ranging from a few minutes to several hours. The dynamic nature of the histone acetylation–deacetylation cycle is readily seen by treating cells with inhibitors of histone deacetylating enzymes, such as the salts of butyric or propionic acid. This results in the progressive accumulation of the more acetylated isoforms, often detectable within a matter of minutes. Because each acetate group added to a histone reduces its net positive charge by one, the acetylated isoforms differ in net charge and so can be separated by electrophoresis (Fig. 4.4). However, each electrophoretic isoform is not necessarily homogeneous but may itself be a mixture of molecules in which the total number of acetate groups is the same but the sites acetylated are different. For example, in the mono-acetylated isoform of H4 in Fig. 4.4, the single acetate could be attached to lysine 5, 8, 12 or 16. The electrophoretic mobility would be the same in each case. As will be discussed in later chapters, which specific lysines are acetylated may be just as significant, in a functional sense, as the overall change in net charge. Acetylation, like all the modifications discussed here, is tightly regulated. Not all the lysines in the exposed N-terminal tails are acetylated, and of those that are, acetylation occurs in a distinctly nonrandom manner. Patterns of acetylation also differ dramatically between organisms. For example, in cuttlefish (a favourite experimental organism with those who like to be at the seaside) the mono-acetylated isoform of histone H4 is acetylated exclusively at lysine 12. As the level of acetylation increases, lysine 5, 16 and 8 are acetylated in an apparently fixed and invariant order (Table 4.2). In mammalian cells too, H4 acetylation is nonrandom, but the order is quite different. Mono-acetylated H4 from human or bovine cells is acetylated almost exclusively at lysine 16. The subsequent order of acetylation is lysine 8, lysine 12 and finally lysine 5, although site usage is more flexible than in cuttlefish (Table 4.2). Because the structure of the nucleosome is essen-

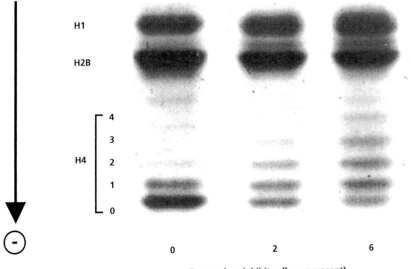

H1

H2B

H4   4
      3
      2
      1
      0

⊖

0   2   6

Deacetylase inhibitor (hours present)

Fig. 4.4 Acetylated isoforms of histone H4 can be resolved by electrophoresis. Histones have a high net positive charge, and if subjected to electrophoresis at low pH, they move towards the cathode. For each lysine that is acetylated, the net positive charge is reduced by one, so the histone moves more slowly towards the cathode. The acetylated isoforms of H4 separate particularly well, and isoforms with 1, 2, 3 or 4 acetate groups can be clearly resolved on gels containing high concentrations of acetic acid, urea and the detergent Triton X-100 (AUT gels). The more acetylated isoforms are very rare in untreated cells, but increase rapidly if cells are exposed to inhibitors of histone deacetylases.

Table 4.2 H4 lysines are acetylated nonrandomly.

	H4 lysines acetylated	
Number of acetates	Mammals	Cuttlefish
1 (mono-acetylated)	16	12
2 (di-acetylated)	16, 12 (or 8)	12, 5
3 (tri-acetylated)	16, 12, 8 (or 5)	12, 5, 16
4 (tetra-acetylated)	16, 12, 8, 5	12, 5, 16, 8

tially the same in mammals and cuttlefish, the differing patterns of acetylation must derive from the properties of the enzymes involved rather than from the structure of the nucleosome.

Protein sequencing has shown that acetylation of histones H3 and H2B from human and bovine cells is also nonrandom, but with a degree of flexibility reminiscent of that seen with H4. It should be emphasized that this type of analysis shows only the *steady-state* level of acetylation at each site for each of the acetylated isoforms. For each isoform this level may be the net result of several different acetylation–deacetylation cycles, possibly occurring at different rates, in different parts of the genome and catalysed by different enzymes.

Methylation

The first sequencing studies of histones H3 and H4 showed that certain lysine residues were methylated on their side-chain ε-amino groups. Methyl-lysine, like acetyl-lysine, is chemically stable. Unlike acetylation, methylation of the lysine ε-amino group does not remove its positive charge, so histones cannot be separated electrophoretically on the basis of differences in methylation. The modification is also complicated by the fact that 1, 2 or 3 methyl groups can be added to each ε-amino group. The methyl donor is the high-energy compound, S-adenosyl methionine (Fig. 4.5).

Methylation occurs most frequently, and possibly exclusively, on histones H3 and H4 in vertebrates. Radiolabelling experiments have shown that the modification generally occurs after chromatin assembly, rather than on newly synthesized histones. Sequencing of purified histones showed that H3 could be methylated at lysines 9 and 27 and H4 at lysine 20. More recent work involving labelling of cultured (HeLa) cells with tritiated S-adenosylmethionine, has shown that H3 lysine 4 is also commonly methylated. Mono- and di-methylated lysines were both detected. In lower eukaryotes, specifically yeast and the ciliated protozoan *Tetrahymena*, only lysine 4 of H3 is methylated, with the mono- and di-methylated forms occurring in roughly equal amounts. Thus, as with acetylation, patterns of methylation seem to differ from one organism to another.

A characteristic that makes *Tetrahymena* such a useful experimental organism is that it has two nuclei per cell. One, the macronucleus, is transcriptionally active and the other, the micronucleus, is not. Both methylated H3 and histone

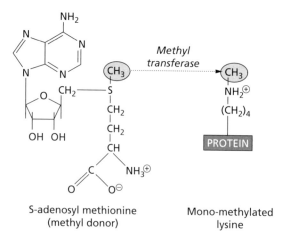

S-adenosyl methionine
(methyl donor)

Mono-methylated
lysine

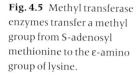

Fig. 4.5 Methyl transferase enzymes transfer a methyl group from S-adenosyl methionine to the ε-amino group of lysine.

methyltransferase enzyme activity are greatly enriched in macronuclear extracts, as are acetylated histones and histone acetyltransferase activity. A link between histone acetylation and methylation has also been made from work with cultured HeLa cells, in which newly methylated H3 was found to be preferentially packaged in nucleosomes containing acetylated H4. In contrast, newly methylated H4 was under-represented in acetylated chromatin. One interpretation of these findings is that both methylation and acetylation play important roles in transcriptionally active chromatin. This remains to be determined, but for now, the most important point is that the two modifications are likely to be interrelated in some functionally significant way. It may be that one modification (e.g. acetylation) predisposes the nucleosome to another (e.g. methylation) and the double-modified (i.e. methylated and acetylated) nucleosomes may have properties that the singly modified versions do not. It is becoming increasingly clear that individual histone modifications should not be considered in isolation. The functional properties of the nucleosome, and higher-order chromatin, are likely to depend on the combination of modifications that it possesses. The types of modification involved, their frequencies and the amino acids involved will all be important.

Nuclear enzymes that are capable of methylating histone lysines have been identified and partially characterized and shown to play important roles in chromatin condensation and gene silencing (Chapters 9 and 10). An enzyme (CARM1) that methylates arginine residues on H3 and other proteins *in vitro* has recently been identified and its gene cloned and sequenced. CARM1 binds to members of the p160 family of coactivators, proteins that are involved in gene activation by nuclear hormone receptors (Chapter 8). A mutant CARM1 protein lacking methyltransferase activity was found to be a much less efficient coactivator than its catalytically active counterpart. However, there is so far no evidence for the *in vivo* methylation of histone arginine residues in higher eukaryotes, even though they have been detected in *Drosophila*.

Phosphorylation

In vivo, proteins can be phosphorylated at two types of amino acid side-chain. Serine, threonine and, rarely but crucially, tyrosine residues can be modified by substituting a phosphate for a hydroxyl group to give an O-phosphate linkage (Fig. 4.6). Lysines, histidines and arginines can be phosphorylated to give N-phosphate linkages. Both types of modification are brought about by specific families of enzymes (the protein kinases) and reversed by others (the protein phosphatases). The phosphate donor can be a nucleotide triphosphate, commonly ATP or GTP or cyclic AMP. These enzymes are involved in numerous cell-signalling pathways, with the attachment and removal of phosphate groups from selected amino acids being used to change the behaviour of the protein in some functionally significant way. As described earlier, protein kinases are

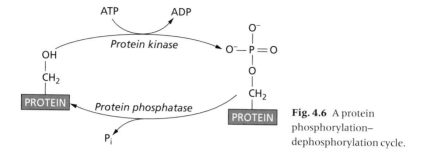

Fig. 4.6 A protein phosphorylation–dephosphorylation cycle.

known to be integral components of the PolII initiation complex and their activity is an essential element in assembly of a functional complex. For example, phosphorylation of the C-terminal domain of a catalytic subunit of PolII is required for transcription initiation and progression. But does phosphorylation of the structural components of chromatin itself, i.e. the histones, also have a role to play?

Although many different protein kinases have been shown to be capable of phosphorylating histones *in vitro* at a variety of amino acid residues, histones are modified *in vivo* only at selected serine and threonine residues. The most extensively phosphorylated histone is the linker histone H1. Five serine and threonine residues in the N- and C-terminal tails can be phosphorylated, with the level of phosphorylation being highest in mitotic cells. Some of this is attributable to the action of growth-associated, cyclin-dependent kinases (CDKs). The role of H1 phosphorylation in cells progressing through mitosis remains a mystery. It has always seemed likely that H1, with its ability to interact with both core and linker DNA, will play a role in the formation of higher-order chromatin structures, but exactly what this role might be is still unclear.

Of the core histones, phosphorylation of H3 has been most extensively studied and provides another intriguing example of how functionally important *combinations* of modifications can be. H3 is phosphorylated exclusively at serine 10 and in two separate circumstances. First, there is a highly conserved and widespread phosphorylation of H3 S10 as cells enter mitosis. Second, there is a much more limited phosphorylation of the same residue in cells stimulated from quiescence to growth by growth factors. What makes the latter particularly interesting is that the phosphorylation is located precisely at those genes whose transcription is activated by growth factors, the so-called Immediate Early genes. This presents us with a paradox: on the one hand, the modification is associated with chromatin condensation and suppression of transcription as cells enter mitosis, and, on the other, it is associated with the transcriptional activation of selected genes. A possible explanation has come from experiments studying, in parallel, the phosphorylation and acetylation status of H3. In growth factor stimulated cells, the small fraction of phosphorylated H3 is associated with highly acetylated chromatin, whereas in mitotic cells it is not. It may be that acetylation predisposes the chromatin to subsequent phosphorylation, possibly by making it a more attractive substrate for the relevant kinase.

The kinases responsible for H3 phosphorylation have still not been identified conclusively, but it seems likely that different enzymes catalyse the mitotic and growth-related phosphorylations. Rsk2 is a strong candidate for the growth-associated phosphorylation. This kinase is absent in patients with the human genetic disease Coffin–Lowry Syndrome. Cells from these patients fail to phosphorylate H3 S10 on growth factor stimulation, but do phosphorylate it when they enter mitosis. Although this might seem conclusive evidence that Rsk2 phosphorylates H3 S10 in growth stimulated cells, but not mitotic cells, there is still the possibility that the role of Rsk2 is to phosphorylate not H3, but the H3 kinase, thereby converting it from an inactive to an active form. Such kinase 'cascades' are very well known in the signalling field. A hypothetical example involving an unspecified histone kinase is illustrated in Fig. 4.7. Absence of either the histone kinase or the histone kinase kinase would prevent histone phosphorylation.

ADP-ribosylation

Histones and other nuclear and cytoplasmic proteins can be modified by attachment of ADP-ribose moieties. The ADP-ribose is derived from nicotinamide adenine dinucleotide (NAD^+), an important coenzyme present throughout the cell in millimolar concentrations and a mediator (with its reduced form NADH) of many oxidation–reduction reactions. The enzyme ADP-ribosyl transferase catalyses the attachment of NAD-derived ADP-ribose to glutamic acid residues (usually), with displacement of the nicotinamide moiety (Fig. 4.8). The modification can be reversed by the enzyme ADP-ribosyl protein lyase. This is all relatively straightforward, but what makes this one of the most complex and

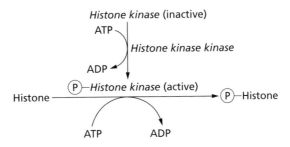

Fig. 4.7 A hypothetical histone kinase cascade. The model proposes that the histone kinase is active only when phosphorylated, so phosphorylation of the histone requires that both the histone kinase itself and the enzyme that phosphorylates it (the histone kinase kinase) must be active. This is a simple protein kinase 'cascade'. It can be extended by proposing that the histone kinase kinase must also be phosphorylated. Such cascades are a common feature of intracellular signalling pathways. The activity of the protein kinases is often counterbalanced by protein phosphatases (as shown in Fig. 4.6).

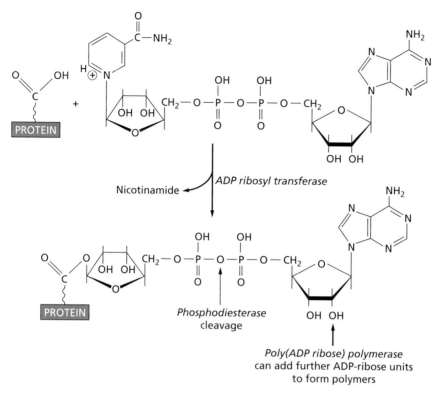

Fig. 4.8 The chemistry of ADP-ribosylation.

variable of all the modifications to which histones are subject is that another en-zyme, poly(ADP-ribose) polymerase, can add additional ADP-ribose units to the original one (Fig. 4.8). Not all mono(ADP-ribose) attachments are extended in this way, but when they are the result can be a polymer, sometimes branched, of over 200 ADP-ribose units. Both phosphodiesterase and glycohydrolase en-zymes can split the polymer.

The activity of poly(ADP-ribose) polymerase is dependent on DNA strand breaks. Treatment of cells with reagents such as dimethyl sulphate (DMS) that cause elevated levels of strand breaks leads to a big increase in the levels of poly ADP-ribosylated histones, but not of the mono ADP-ribosylated forms. Two-dimensional electrophoresis, separating histones by charge in the first direc-tion, and size in the second, resolved dozens of different ADP-ribosylated isoforms of H1, H2B and other core histones in DMS-treated cells. However, while the multiple isoforms are readily detectable in histones extracted from cells grown in radiolabelled NAD^+, they are not detectable by Coomassie Blue staining. They must therefore comprise only a small proportion (less than 5%) of total histone. It is also worth noting that poly(ADP-ribose), present in free form in DMS-treated cells, binds strongly to the tail domains of all core histones. Such binding has interesting functional implications, but may also contribute to the labelling detected on 2D gels. Only H2B has a glutamic acid residue located in a position likely to be accessible to the ADP-ribosyl transferase enzymes in chromatin.

Poly(ADP-ribose) polymerase has a DNA-binding domain with two zinc fingers (Fig. 1.6) and binds adjacent to single- and double-strand breaks, so the modified histones are presumably localized to sites of DNA repair. They may function to facilitate chromatin remodelling during the repair process. The N-terminal tails of histones H3 and H4 have as great an affinity for poly(ADP-ribose) as for DNA, so the presence of this polymer on H2B, or other proteins, could help dissociate H3 and H4 from DNA. The polymers could also act as chaperones, protecting the charged N-terminal tails while the hydrophobic globular domains of the core histones reassociate and organize DNA into a new nucleosome.

Ubiquitination

The 76 amino acid peptide ubiquitin is a good candidate for the world's most highly conserved protein. It is found throughout the animal and plant kingdoms and in prokaryotes. It is usually found attached to other proteins, either as a single unit or as a polymeric chain. The modification requires two or three enzymes: a ubiquitin-activating enzyme E1, a ubiquitin-conjugating (or -carrier) enzyme E2, and sometimes a ubiquitin–protein ligase E3. Ubiquitin is attached to the ε-amino group of selected lysines by an unusual 'isopeptide' linkage (Fig. 4.9). Additional ubiquitin peptides can be attached to the first one by the same type of linkage, generating a poly-ubiquitinated protein. The major role of ubiquitination is in targeting proteins for degradation by intracellular protein-digesting factories, the proteosomes, although this is unlikely to be its major purpose on ubiquitinated histones in chromatin.

The most frequently ubiquitinated histones are H2A (at lysine 119) and H2B (at lysine 120). Ubiquitinated H3 has been detected in elongating spermatids. In higher eukaryotes, 10% of H2A and 1–2% of H2B molecules are typically present as either mono- or poly-ubiquitinated isoforms, although the levels vary depending on the developmental stage or state of growth of the cells. Ubiquitinated H2B, and to a lesser extent H2A, have been associated with ongoing

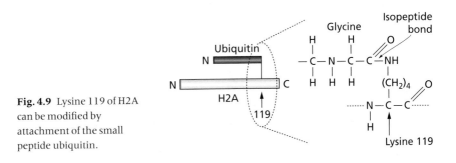

Fig. 4.9 Lysine 119 of H2A can be modified by attachment of the small peptide ubiquitin.

transcription (i.e. they seem to be required to keep transcription going rather than just to get it started). The high-resolution crystal structure of the nucleosome core particle shows that lysine 120 of H2B is buried and likely to be inaccessible to modifying enzymes. Perhaps this changes during transcription, allowing attachment of ubiquitin, which may then maintain an altered structure favourable to continuing transcription. It is hard to see how the presence of such a bulky attachment to an internal residue could fail to alter the structure of the nucleosome in one way or another.

Histone variants

In higher eukaryotes, histone genes are present in multiple copies. Usually, the genes for the five different histones are present in a cluster, which is then repeated many times. Only in this way can the cell provide the vast numbers of histones needed to package newly replicated DNA during S-phase. Expression of histone genes is at its highest during S-phase, with only residual expression during the rest of the cell cycle, providing sufficient histone to cope with the demands of DNA repair and associated chromatin remodelling activities. Most copies of the genes encoding any given histone are the same. However, for all the histones apart from H4, there are variant genes encoding histones with differences in amino acid sequence.

In some cases these differences are relatively subtle, comprising amino acid substitutions that have little or no apparent effect on the properties or function of the histone. For example, there are three H3 variants in mammals, H3.1, H3.2 and H3.3. H3.2 differs from H3.1 by a single amino acid substitution (serine for cysteine at position 96), while H3.3 has two additional substitutions. These substitutions are enough to cause major changes in the mobility of the three variants after electrophoresis on acid/Triton/urea gels, but they seem to behave equivalently in most experimental systems (e.g. in the frequency with which they are acetylated or methylated). However, whereas H3.1 and H3.2, like most histones, are synthesized only during S-phase, H3.3 is synthesized throughout the cell cycle, so the variants are not equivalent in all respects.

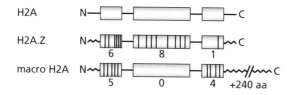

Fig. 4.10 Non-allelic variants of H2A. H2A.Z and macroH2A are encoded by different genes to H2A itself. The N- and C-terminal tail domains of the variants are quite different from those of H2A, and the C-terminal domain of macroH2A is greatly extended. The variants also have amino acid substitutions in the helices of the histone fold domain (vertical lines). The number of substitutions in each helix is indicated.

In other cases, the differences in primary structure between variants are much more radical. For example, the human H2A variant H2A.Z differs from the common variant H2A.1 by 15 amino acid substitutions in the three helical domains along with numerous substitutions, insertions and deletions in the N- and C-terminal and loop domains (Fig. 4.10). The same variant is found in *Drosophila* and is closely homologous to the human version. Presumably, the H2A.Z variant arose at an early stage in evolution and has been conserved, suggesting that it has a specific and fundamental functional role. Its role may be in regulation of transcription, as the H2A.Z homologue in *Tetrahymena* has been shown to associate preferentially with transcriptionally competent chromatin in the active macronucleus. A second H2A variant, macroH2A, differs to an even greater extent from H2A.1. In addition to an array of amino acid substitutions, it has over 200 additional amino acids added to its C-terminal tail (Fig. 4.10). MacroH2A has been shown to associate specifically with the *inactive* (i.e. transcriptionally silent and heterochromatic) X chromosome in female mammals (Chapter 12), suggesting a role in chromatin condensation and gene silencing.

A tail extension, this time at the N-terminal end, is also seen in a variant of H2B found only in sperm. The 21 amino acid extension interacts with linker DNA and may be involved in the extreme condensation of DNA in the sperm nucleus. Developmental, or cell type-specific expression of histone variants is not uncommon. For example, in sea urchins there are four H2B variants, in addition to sperm H2B and seven H2A variants (but none for H3 and H4). Levels of expression of all these variants change in characteristic ways through sea urchin development.

Further reading

Histone tails

Hansen, J. C., Tse, C. & Wolffe, A. P. (1998) Structure and function of the core histone N-termini: more than meets the eye. *Biochemistry*, **37:** 17637–17641.

Luger, K. & Richmond, T. J. (1998) The histone tails of the nucleosome. *Curr. Opin. Genet. Dev.*, **8:** 140–146.

Histone modifications

Spencer, V. A. & Davie, J. R. (1999) Role of covalent modifications of histones in regulating gene expression. *Gene*, **240:** 1–12.

Acetylation

Kouzarides, T. (2000) Acetylation: a regulatory modification to rival phosphorylation. *EMBO J.*, **19:** 1176–1179.

Morales, V. & Richard-Foy, H. (2000) Role of histone N-terminal tails and their acetylation in nucleosome dynamics. *Mol. Cell Biol.*, **20:** 7230–7237.

Turner, B. M. (2000) Histone acetylation and an epigenetic code. *BioEssays*, **22:** 836–845.

Methylation

Annunziato, A. T., Eason, M. B. & Perry, C. A. (1995) Relationship between methylation and acetylation of arginine-rich histones in cycling and arrested HeLa cells. *Biochemistry*, **34:** 2916–2924.

Chen, D., Ma, H., Hong, H. *et al*. (1999) Regulation of transcription by a protein methyltransferase. *Science*, **284:** 2174–2177.

Rea, S., Eisenhaber, F., O'Carroll, D., Strahl, B. D. *et al*. (2000) Regulation of chromatin structure by site-specific histone H3 methyltransferases. *Nature*, **406:** 593–599.

Strahl, B. D., Ohba, R., Cook, R. G. & Allis, C. D. (1999) Methylation of histone H3 at lysine 4 is highly conserved and correlates with transcriptionally active nuclei in *Tetrahymena*. *Proc. Natl. Acad. Sci., USA*, **96:** 14967–14972.

Phosphorylation

Hendzel, M. J., Wei, Y., Mancini, M. A. *et al*. (1997) Mitosis-specific phosphorylation of histone H3 initiates primarily within centric heterochromatin during G_2 and spreads in an ordered fashion coincident with mitotic chromosome condensation. *Chromosoma*, **106:** 348–360.

Hsu, J.-Y., Sun, Z.-W., Li, X., Reuben, M. *et al*. (2000) Mitotic phosphorylation of histone H3 is governed by Ip1/aurora kinase and Glc7/PP1 phosphatase in budding yeast and nematodes. *Cell*, **102:** 279–291.

Thomson, S., Mahadevan, L. C. & Clayton, A. L. (1999) MAP kinase-mediated signalling to nucleosomes and immediate-early gene induction. *Sem. Cell Dev. Biol.*, **10:** 205–214.

ADP-ribosylation

Althaus, F. R. (1992) Poly ADP-ribosylation: a histone shuttle mechanism in DNA excision repair. *J. Cell Sci.*, **102:** 663–670.

Jacobson, M. K. & Jacobsen, E. L. (1999) Discovering new ADP-ribose polymer cycles: protecting the genome and more. *TIBS*, **24:** 415–417.

Ubiquitination

Baarends, W. M., Hoogerbrugge, J. W., Roest, H. P. *et al*. (1999) Histone ubiquitination and chromatin remodeling in mouse spermatogenesis. *Dev. Biol.*, **207:** 322–333.

Huang, H. H., Kahana, A., Gottschling, D. E., Prakash, L. & Liebman, S. W. (1997) The ubiquitin-conjugating enzyme Rad6 (Ubc2) is required for silencing in *Saccharomyces cerevisiae*. *Mol. Cell. Biol.*, **17:** 6693–6699.

Moazad, D. & Johnson, A. D. (1996) A deubiquitinating enzyme interacts with SIR4 and regulates silencing in *Saccharomyces cerevisiae*. *Cell*, **86:** 667–677.

Chapter 5: Higher-Order Chromatin Structures and Nuclear Organization

Introduction

The problem with which the last chapter opened was that of packaging an enormously long length of DNA into the nucleus in such a way that it could carry out its various functions. After defining the basic unit of chromatin structure, the nucleosome core particle, some of its properties were outlined, particularly those that are relevant to a possible role in regulating gene expression. However, the nucleosome is only the first stage in packaging DNA into the cell nucleus. In order to appreciate just how far beyond the nucleosome packaging needs to progress, the figures should be quickly restated. There is about 2 m of DNA in the nucleus of every human cell. In contrast, the total length of all 46 human chromosomes at metaphase is about 200 μm or 2×10^{-4} m. (At least, this is the length of all 46 chromosomes fixed and spread on glass slides.) So, the total reduction in the length of the genome (sometimes referred to as the packaging ratio) as cells go through the metaphase stage of mitosis is about 10 000-fold. The packing ratio will be less in the interphase nucleus, where the chromosomes, overall, are much less condensed, but is still likely to be something of the order of 250-fold. The supercoiling of DNA in the nucleosome reduces its overall length by only about seven-fold, which does not even come close to what is required. There must be other levels of DNA folding that bridge the yawning gap between the nucleosome and the metaphase chromosome. Experimentation to find out what these might be originally centred around the techniques that had proved so successful in defining the structure of the nucleosome, namely electron microscopy and X-ray crystallography. Unfortunately, the technical problems encountered with the nucleosome appear trivial when compared with those involved in analysing even larger structures. However, analysis by electron microscopy of nuclei lysed under carefully controlled conditions showed that a fibre with a diameter of 30 nm was consistently present (examples are shown in Fig. 5.1).

The 30 nm fibre

Model building has proved particularly useful in devising possible higher-order chromatin structures. As always, it is best to start off with simple structures and work up, and the logical first step is to build a di-nucleosome. This rather modest step proves to be quite informative, because the di-nucleosome places some serious constraints on likely structures for the 30 nm fibre. The path of DNA in a di-nucleosome is such that as the DNA leaves the second nucleosome, it is

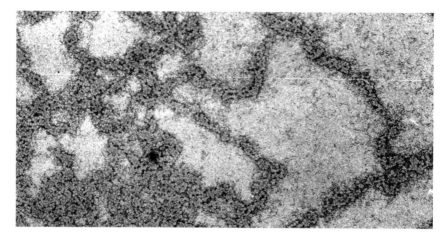

Fig. 5.1 An electron micrograph showing chromatin fibres of around 30 nm diameter and apparently made up of coiled nucleosomes. Such images define the diameter of the fibre, but can provide little evidence for or against the various structural models outlined in the text (photograph provided by Barbara Hamkalo).

directed, more or less, towards the point at which it entered the first nucleosome, i.e. it has turned back on itself. (Follow the path of the DNA linking nucleosomes 1 and 2 in Fig. 5.2.) This behaviour is determined by the geometry of the nucleosome core particle and is true whether the two nucleosomes are the same way up (Fig. 5.2a) or in opposite orientations (Fig. 5.2b). This DNA path clearly does not favour a completely linear, beads-on-a-string structure, but it can be accommodated by a fibre in which the DNA follows a 'zig-zag', or 'sawtooth' path, with alternate nucleosomes on one side or the other (Fig. 5.2). The structure is essentially the same whether alternate nucleosomes are the same way up (in which case the linker DNA path crosses itself once per nucleosome, Fig. 5.2a) or inverted (in which case linker DNA cross-overs occur only every other nucleosome, Fig. 5.2b). Zig-zag arrangements of nucleosomes in extended chromatin fibres are frequently been seen in EM pictures of spread chromatin. Examples can be found in Fig. 3.2.

Models have been proposed for the 30 nm fibre that attempt to accommodate the preferred path of DNA through the di-nucleosome. The models differ in detail, but most can be categorized into one of two general types. One of these, the twisted-ribbon model, is shown in Fig. 5.3. It is assembled by taking the zig-zag chromatin fibre and wrapping it around an imaginary cylinder to form a 30 nm diameter helix. This arrangement creates two parallel helices containing either odd- or even-numbered nucleosomes. The version illustrated in Fig. 5.3 has 18 nucleosomes per turn of the double helix, a helical repeat length (pitch) of about 32 nm and a hollow central core. The model nicely accommodates the preferred path of linker DNA and has a conceptually simple assembly mechanism, namely coiling of the zig-zag chromatin fibre. It has the disadvantage of placing both linker DNA and linker histones on the outside of the fibre. Most

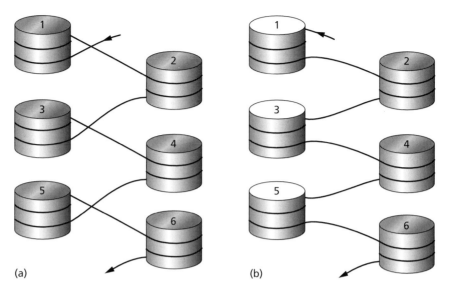

Fig. 5.2 The path of DNA between nucleosomes. In (a) the nucleosomes in the fibre are all the same way up, whereas in (b) alternate nucleosomes are inverted. The two structures differ in the number of times the DNA path crosses over itself in a run of nucleosomes.

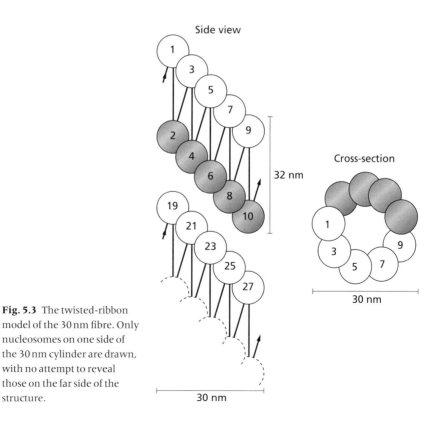

Fig. 5.3 The twisted-ribbon model of the 30 nm fibre. Only nucleosomes on one side of the 30 nm cylinder are drawn, with no attempt to reveal those on the far side of the structure.

experimental evidence suggests that both are primarily internal, where they are protected from nucleases and proteases respectively.

The 'crossed-linker' model of the 30 nm fibre is superficially similar to the twisted ribbon, in that it contains two parallel helices of alternating odd- and even-numbered nucleosomes with about 18 nucleosomes per turn of the double helix (Fig. 5.4). However, the path of the linker DNA is quite different. Nucleosomes that are next to one another in the zig-zag fibre take up positions on opposite sides of the helix and the linker DNA joining them bridges the helical core (Fig. 5.4). A prediction of the model is that the width of the helix will be proportional to the length of the linker DNA, and there is some evidence that this is indeed the case. Both linker DNA and H1 are sheltered inside the fibre. It is not immediately obvious from this model how the 30 nm fibre can be assembled from the linear zig-zag fibre, but the problem is not as great as it may first seem. Fig. 5.5 (left side) shows six nucleosomes as a zig-zag fibre and, on the right, the same six nucleosomes organized as part of a crossed-linker fibre. The transition can be made by compressing the filament from the top and applying a clockwise twist. (This is not a suggestion as to how the 30 nm fibre might be assembled *in vivo*, just a way of visualizing how a transition between the two structures might be achieved.)

A third model for the 30 nm fibre is the most straightforward and is sometimes presented as the definitive structure, although the evidence in its favour is not conclusive. It is a simple solenoid with about six nucleosomes per turn, a pitch of about 11 nm and the linker DNA and H1 tucked away on the inside (Fig. 5.6). Nucleosomes that are side by side in the linear fibre remain side by side in the solenoid. This model is supported by a significant amount of experimental data, not least X-ray diffraction data from the Cambridge group who were so successful in defining the structure of the nucleosome itself. However, it makes few concessions to the preferred path of the linker DNA. As illustrated in the

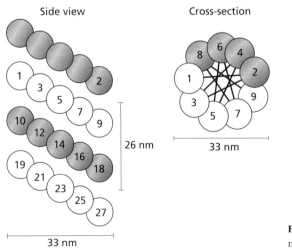

Fig. 5.4 The crossed-linker model of the 30 nm fibre.

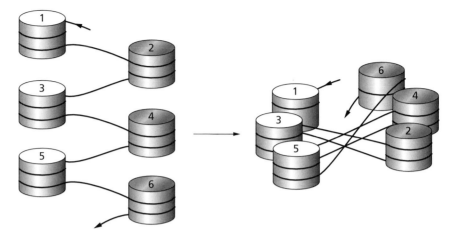

Fig. 5.5 Diagram showing how a linear arrangement of nucleosomes might be collapsed to form a crossed-linker fibre.

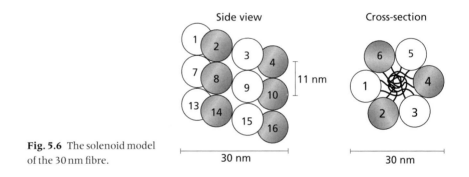

Fig. 5.6 The solenoid model of the 30 nm fibre.

hypothetical structure in Fig. 5.7, the linker DNA must curve sharply between adjacent nucleosomes. It is easy enough to deal with this when representing the path of the DNA as a thin line, as in the diagram. In reality, the DNA is a much wider structure with significant rigidity.

The 30 nm fibre has been extensively studied by various experimental approaches, including electron microscopy, X-ray diffraction, circular dichroism and hydrodynamic measurements (centrifugation, etc.). Unfortunately, all these approaches suffer from limitations and potential artefacts that prevent them from providing definitive data that would allow us to decide on the detailed structure of the 30 nm fibre. We cannot even be sure whether the nucleosomes are organized in left-handed or right-handed helices (left-handed versions are shown in all the diagrams used here), or whether the path of the linker DNA crosses once or twice between nucleosomes (Fig. 5.2). We do not know how individual nucleosomes are orientated within the fibre or how they interact with one another, although there is strong evidence to show that both

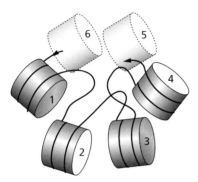

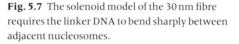

Fig. 5.7 The solenoid model of the 30 nm fibre requires the linker DNA to bend sharply between adjacent nucleosomes.

the linker histone H1 and the N-terminal tails of the core histones are necessary for assembly.

In evaluating the various models, the twisted-ribbon model is probably the least consistent with existing experimental evidence, but the crossed-linker and the solenoid models (and their variants) both have strong adherents. It is an unfortunate fact that, until techniques for imaging and dissecting large and flexible structures *in vivo* are improved, the structure of the 30 nm fibre will remain uncertain and controversial.

DNA loops

Some imaginative experiments carried out in the 1960s gave an important insight into the organization of DNA itself within the nuclei of eukaryotic cells and have provided important clues about chromatin organization beyond the 30 nm fibre. If nuclei are prepared from a cultured cell line (human HeLa cells are commonly used) and then extracted with 2 M NaCl, all the histones are removed, along with many other proteins, although a recognizable nucleus-like structure (referred to as a nucleoid) remains. Provided suitable precautions are taken to inhibit endogenous nucleases, the DNA is also preserved and can be visualized with fluorescent dyes as a halo surrounding what remains of the nucleus. (An electron micrograph showing the central region of a typical nucleoid is shown in Fig. 5.8; DNA fibres spread far beyond the region shown.) This is not unexpected: DNA is after all an extremely long polymer, and this alone might be enough to keep it tangled up in the nuclear remnant. But a surprising result came when nucleoids were exposed to dyes such as ethidium that are able to slide in between individual base pairs and alter the conformation of DNA. Such intercalating dyes force the base pairs apart and cause the DNA to unwind (i.e. there is an increase in the number of base pairs per turn, as illustrated in Fig. 3.9). Linear DNA can accommodate this unwinding simply by rotating its free ends and increasing in length. However, when nucleoids were exposed to high concentrations of ethidium, the halo of DNA, rather than expanding, became *more compact*. This ethidium-induced compaction of DNA

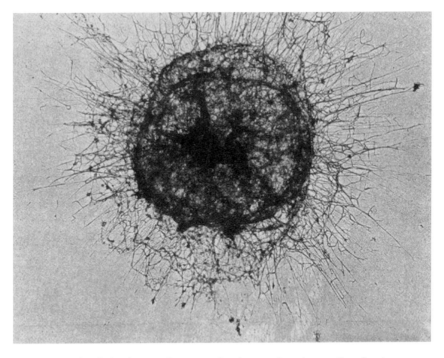

Fig. 5.8 A nucleoid; the electron-dense complex that remains when a cell nucleus is subjected to high salt extraction. The DNA spilling out of the nucleoid forms a wide halo, with a diameter several times that of the nucleoid (about 15 μm). (From McCready *et al.* 1979, *J. Cell. Sci.*, **39**, 53.)

was detected by two different techniques. One involved measuring the diameter of the DNA halo extending from dye-stained nucleoids under the fluorescence microscope, while the other measured their sedimentation rate during centrifugation through sucrose gradients. Both gave the same puzzling result.

The behaviour of the DNA in nucleoids can be explained in terms of its higher-order folding, a behaviour that comes under the general heading of 'supercoiling'. We have already encountered supercoiling when describing the path of DNA around the nucleosome, and given that virtually all eukaryotic DNA is packaged as nucleosomes, the presence of supercoiling is not, in itself, surprising. What is surprising about these results is that they suggest strongly that *additional* DNA supercoils are formed at high concentrations of intercalating dyes, a behaviour that had previously been seen only in the *closed circular* DNAs found in bacteria and viruses. For a closed circular DNA molecule, general unwinding is not an option. (*Local* unwinding might occur, but would necessarily cause *overwinding* in another place; there can be no *overall* change in base pairs per helical turn.) Closed circular DNA must find some other way of accommodating the stresses caused by the intercalating dye. It does this by *coiling the DNA helix itself*, a property referred to as supercoiling. The process is shown in Fig. 5.9. Supercoiling can certainly explain the contraction of the DNA halo in

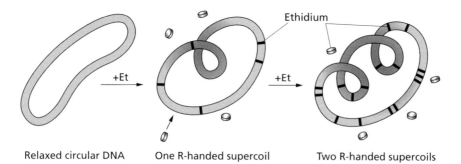

Fig. 5.9 Intercalating dyes such as ethidium (Et) induce supercoiling of circular DNA.

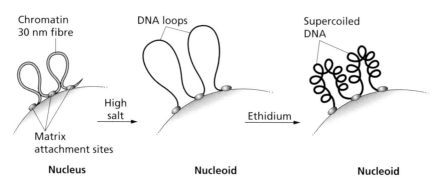

Fig. 5.10 DNA loops attached to a nuclear matrix behave like circular DNA by supercoiling in the presence of high concentrations of ethidium.

nucleoids, but human DNA, and eukaryotic DNA in general, is manifestly not circular. The interpretation put forward to resolve this paradox was that DNA in the nucleus is arranged in the form of loops, with the bases of the loops attached to structural elements in the nucleus. Attachment of the loops in this way will cause the DNA to respond to intercalating dyes in the same way as closed circular DNA (Fig. 5.10).

The concept that DNA in human cells is organized as loops was validated by a completely different experimental approach. If human metaphase chromosomes are depleted of histones (e.g. by treatment with 2 M NaCl) and then spread, at low ionic strength, on carbon-coated electron microscope grids, loops of DNA can be seen emanating from a central core structure designated the chromosome scaffold (Fig. 5.11; needless to say, the experiments are not as simple as this brief description might suggest and represent something of a technical *tour de force*). Because the loops were visualized directly, they could be measured. They varied in size around a mean of about 70 μm, corresponding to 200 kb of B-form DNA. Some other preparation procedures gave rosettes rather than longer loops, but the message was essentially the same — DNA is attached, periodically, to an underlying structure in such a way as to form closed loops.

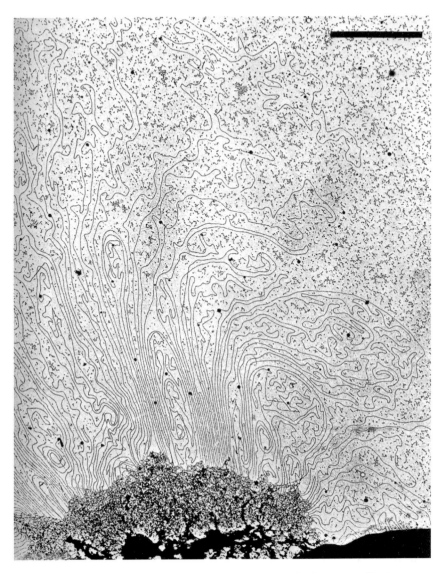

Fig. 5.11 DNA in a human metaphase chromosome is arranged as loops extending from a central electron-dense core, the chromosome scaffold. For most loops, the two ends emerge from the core region (dark area) alongside one another. When the lengths of loops along one side of a small chromatid were measured, values ranged from less than 5 μm up to almost 50 μm. Most genomic DNA is found in loops between 10 and 30 μm long (30–90 kb). Bar = 2 μm. (From Paulson & Laemmli, 1977.)

The nuclear matrix and chromosome scaffolds

If DNA within the nucleus is indeed organized as loops, as indicated by electron microscopy and treatment with intercalating dyes, then some sort of structure must be present that tethers the bases of the loops. The question of what this structure is has exercised many members of the scientific community for many years and has generated more than its fair share of controversy.

As mentioned earlier, if nuclei are exposed to buffers containing ionic detergents and high salt concentrations, a structure remains that preserves the appearance, if only in outline, of the original nucleus. Removal of nucleic acids by prolonged digestion with nucleases fails to destroy this structure, although destruction of RNA (but not DNA) does alter its composition and microscopical appearance. The structure that remains after these harsh extraction procedures is generally referred to as the nuclear matrix. The appearance and protein composition of the nuclear matrix varies according to the extraction procedures used. (An example is given in Fig. 5.12.) It has often been suggested that the residual structure is a consequence primarily of aggregation of nuclear components induced by the extraction procedures themselves. It has certainly proved difficult or impossible, in the great majority of cell types, to visualize convincingly, within intact or mildly extracted nuclei, the sort of filamentous network that has been hypothesized on the basis of nuclear extraction procedures, i.e. something resembling the cytoplasmic filament network, the cytoskeleton. After several years of intensive research into the nuclear matrix and related structures, only one proteinaceous network has been reliably identified and well-characterized within eukaryotic nuclei. This network, the nuclear lamina, is made up of three proteins (lamins A, B and C), one of which (lamin A) has structural properties that closely resemble those of major components of the cytoskeleton, the intermediate filaments. The nuclear lamina lines the inner surface of the nuclear envelope but does not extend, as far as we know, into the body of the nucleus. The lamins are undoubtedly important structural components and may well have a role in regulating nuclear functions, but they are unlikely to be responsible for tethering DNA loops *throughout* the nucleus.

The nuclear lamins are usually prominent protein bands when nuclear matrix preparations are resolved by electrophoresis on polyacrylamide gels, but they are found amongst a large number of different, electrophoretically distinct proteins. In some preparations, the majority of these are proteins known to be

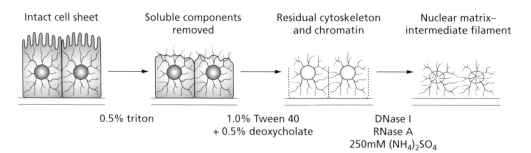

Intact cell sheet	Soluble components removed	Residual cytoskeleton and chromatin	Nuclear matrix– intermediate filament
	0.5% triton	1.0% Tween 40 + 0.5% deoxycholate	DNase I RNase A 250mM $(NH_4)_2SO_4$

Fig. 5.12 A progressive extraction protocol for preparing nuclear matrix–intermediate filament complexes from an epithelial cell sheet. There are many variations on this basic theme. (Based on the procedure of Fey *et al.*, 1973.)

associated with RNA components of the nucleus and involved in RNA processing. These findings raise the possibility that the nuclear matrix, *in vivo*, comprises a rather loose (and/or dynamic) association of protein and RNA components. Such a model would certainly explain the difficulty in isolating matrices as separate and discrete structures. This dynamic model also raises the question of how closely matrices isolated *in vitro* can possibly resemble the *in vivo* structure. They cannot be an exact replica in the way that carefully prepared preparations of the comparatively stable cytoskeleton can be. To distinguish between a reasonable facsimile and a completely artefactual aggregation of components induced by the preparation procedure remains a major problem.

The first, spectacular, electron microscope pictures of DNA loops in metaphase chromosomes showed a 'scaffold' linking the bases of the loops and forming a structure with the general morphology of a metaphase chromosome. It was possible to imagine that this scaffold was a proteinaceous template on which the DNA was assembled and by which the structure of the condensed chromosome was determined. As in the case of the interphase nuclear matrix, components of the chromosome scaffold seem to have a tendency to aggregate and the original observations suffered from the same (inevitable) criticisms that were aimed at nuclear matrix preparations, namely that the central scaffold itself was a preparative artefact. However, most people would accept that the bases of the DNA loops must be attached to *something*. Isolation of metaphase chromosome scaffolds and examination of their protein content reveals that, in contrast to nuclear matrices, there are just two major proteins, named SC-I and SC-II. SC-I was quickly shown to be topoisomerase II (topoII), a remarkable enzyme with the ability to disentangle interlinked DNA strands by a cutting and rejoining mechanism. Such an activity is necessary for separation of newly replicated DNA strands during S-phase. TopoII is also intimately involved in chromosome condensation prior to mitosis. Not surprisingly, it is an essential enzyme in all eukaryotes. The second scaffold protein, SC-II, has been identified as a member of a fascinating protein family, the SMC proteins. SMC proteins were first identified in the yeast *Saccharomyces cerevisiae* (as usual, by mutation studies) as proteins essential for the correct condensation and segregation of chromosomes at mitosis (the full name of the genes involved is 'Stability of MiniChromosomes'). Subsequently, related proteins (with similar functions) were found in chickens, amphibians (*Xenopus laevis*) and nematode worms (*Caenorhabditis elegans*) amongst others. The structure of SMC proteins (Fig. 5.13) suggests that they have the ability to bind DNA (through a helix–loop–helix domain), to hydrolyse nucleotide triphosphates (through a domain resembling that found in RNA helicases and transporter proteins) and to associate with other family members to form heterodimers. Overall, the structure of SMC proteins resembles that of motor proteins such as kinesin, a protein that is attached to the cytoskeleton (specifically to microtubules) and is involved in the movement of various subcellular components. In other words, the properties of SC-I and SC-II are exactly what one requires of proteins involved in the manipulation and movement of large subcellular structures (i.e. chromosomes). Sig-

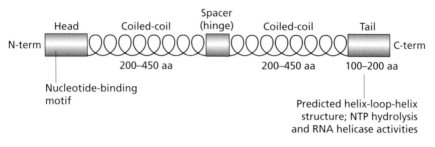

Fig. 5.13 The domain structure of SMC proteins.

nificantly, when antibodies to SC-I and SC-II are used to immunostain metaphase chromosomes prepared under much milder conditions than those used for electron microscopy, they reveal a core running down the centre of each chromatid.

The metaphase chromosome scaffold is apparently a highly conserved structure, at least in terms of its major proteins. However, it would be wrong to oversimplify the situation. SC-I (topoII) and SC-II are proteins that usually remain associated with chromosomal DNA after stringent extraction procedures. Other, functionally essential components, may well be present initially but are lost during preparation. Indeed, SC-I and SC-II (or their equivalents) alone are not sufficient for chromosome condensation and segregation. Alternative approaches have revealed other components, some of which are also members of the SMC protein family. For example, condensation of DNA *in vitro* with extracts from *Xenopus* eggs requires topoII and two SMC proteins, named XCAP-C and XCAP-E. (Note that in these chromosomes, immunostaining shows that the XCAP proteins form a clearly defined chromosome core whereas topoII is distributed throughout the chromosome.) In addition, mutation studies in yeast have revealed several genes that are necessary for chromosome condensation and behaviour through mitosis. It is not surprising to find that such a complex process requires several different proteins and it is important to avoid labels that limit the functional significance of specific components or oversimplify complex situations. SC-I and SC-II clearly have both enzymatic and structural roles. They are components of something termed the metaphase chromosome scaffold, although their distributions may differ depending on the circumstances, but they almost certainly play other roles in the cell and other proteins are certainly associated with them as part of the scaffold *in vivo*. Finally, at the risk of being repetitive, it is important to bear in mind, as we switch from one model system to another, that different organisms may use even highly conserved proteins in different ways. Valuable though model systems are, they always carry the risk that some results at least are organism specific.

At this stage it is important to note that chromatin condensation is not only essential for progression through mitosis but, as discussed at length in later chapters, is also used as a means of regulating gene expression. With this in mind, two recent findings are particularly interesting. First, the two proteins SC-I and SC-II copurify with a protein complex, designated UB2, that contains DNA-binding factors and is likely to be involved in transcriptional regulation. Second, in the nematode *C. elegans*, the protein MIX-1, a member of the SMC family, associates specifically with the interphase X chromosome in XX (hermaphrodite) animals, but not in XY males. It is part of a protein complex known to be involved in the transcriptional down-regulation of X-linked genes in XX animals and is an integral part of the dosage compensation mechanism in *C. elegans*. This is discussed in more detail in Chapter 12. What is remarkable about MIX-1, is that in interphase cells, it is associated only with the X chromosome in XX animals, while in metaphase it is present on *all* chromosomes in both XX and XY animals and has been shown to have a role in metaphase chromosome condensation. Thus, the same protein is involved in gene regulation (as part of the dosage compensation mechanism) and chromosome condensation at mitosis.

Scaffold/matrix associated regions (SARs and MARs)

Nuclear matrix or chromosome scaffold preparations always contain a small amount of DNA that has remained associated through the various extraction and nuclease digestion procedures. This DNA is not a random cross-section of the genome, as would be expected if its association was the result of nonspecific, artefactual attachment, but instead contains a reproducible set of DNA fragments. (The preparation of these fragments begins with a nuclear matrix preparation of the sort outlined in Fig. 5.12, followed by digestion with proteases to release DNA fragments that have been *protected* from the DNaseI digestion step by their association with matrix proteins.) In general, the same DNA fragments seem to be found associated with both interphase nuclear matrix preparations and with metaphase chromosome scaffolds. They are referred to as either matrix associated regions (MARs) or scaffold associated regions (SARs). The two terms are usually interchangeable and I will refer to them as MARs.

Any experiments involving the nuclear matrix are fraught with complications, and the identification of matrix-associated DNA is no exception. The simple fact of finding particular DNA sequences associated with matrix preparations is not, in itself, sufficient to conclude either that the sequence is a 'genuine' MAR, or that it is really matrix associated *in vivo*. When DNA fragments isolated from nuclear matrix preparations are tested for their ability to bind to the DNA-free matrix *in vitro*, only a few show any specific binding ability. Why is this? It is likely that some of the isolated fragments are simply tangled up in the matrix nonspecifically. But others, perhaps the majority, may be there because they were being transcribed when the matrix was isolated. There is evidence to suggest that the transcription apparatus is associated, in some way that is still not understood, with the matrix. These transcribing sequences have no intrinsic matrix-binding ability, they just happened to have been associated with a transcription complex when it was brought to a halt by the matrix preparation procedure. In order to detect those sequences that have the ability to bind *directly* to the matrix, it is necessary to test matrix-associated DNA further by carrying out *in vitro* binding assays using cloned, radiolabelled DNA fragments and DNA-free matrix preparations (prolonged DNaseI treatment and strong detergents can remove most matrix DNA). DNA sequences that are both retained in nuclear matrix preparations *and* show high affinity binding in *in vitro* assays are good candidates for genuine MARs.

Isolated MAR fragments are between 300 and 1000 bp in size and each one contains several sites of DNA–scaffold interaction. MARs are AT-rich, but there are no *simple* sequence motifs that serve to identify them. When the binding of radiolabelled MAR fragments to scaffold/matrix preparations is tested quantitatively, it is found to be saturable, i.e. there is a limited number of available binding sites. It is not known to which specific matrix proteins the MAR fragments bind. Where long, contiguous stretches of genome have been screened, the distances between MARs have ranged from less than 3 kb to 140 kb, i.e. close to the range of loop sizes measured by electron microscopy.

A remaining problem is that, even if a sequence is shown to be matrix-associated and have high-affinity binding *in vitro*, it still cannot be assumed that it is matrix-associated *in vivo*. Again, the reason is technical. The procedures used to isolate matrix preparations are such that there is often a period, albeit transient, when naked DNA fragments are exposed to matrix proteins and may bind to them. For example, sequences that bind strongly to the enzyme topoII, a major matrix protein, are frequent in genomic DNA. These sequences are often found in matrix preparations and some are likely to be genuinely matrix-associated *in vivo*. However, others may become matrix associated only during the preparation procedure. It is likely that the binding to the matrix of at least some MARs and topoII binding elements is prevented, or regulated, through their packaging as chromatin.

One of the best characterized MARs is that which separates the histone gene clusters in *Drosophila melanogaster*. As in all higher eukaryotes, the histone genes in *Drosophila* are clustered and present in multiple copies. As shown in Fig. 5.14, each cluster is separated by a MAR. The MAR is contained in a DNA fragment consisting of two protein binding regions, each of about 200 bp, separated by a 100 bp spacer. It is also rich in sequences that favour topoII binding. The location of the MAR suggests that the 100 or so histone gene clusters in *Drosophila* are organized as a series of small (5 kb) loops. However, as noted earlier, it is difficult to demonstrate conclusively whether or not MARs are *actually* matrix-associated *in vivo*. The location of the histone MARs is certainly suggestive, but whether the series of loop domains shown in Fig. 5.14 is in fact adopted *in vivo* remains to be proven.

It is interesting to consider what the functional significance of MARs might be. They are sometimes thought to be a part of 'boundary elements', regions of

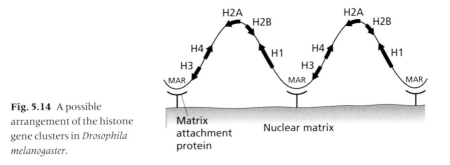

Fig. 5.14 A possible arrangement of the histone gene clusters in *Drosophila melanogaster*.

DNA that define the ends of functional genomic domains and that (i) prevent the genes within them from being influenced by external control elements and (ii) prevent elements within the domain from regulating genes outside it. This is discussed at greater length in later chapters. However, it is not clear what there is to be gained from isolating one cluster of histone genes from the next. They are all subject to the same, cell-cycle-dependent regulation and are, as far as we know, coordinately expressed. It may be that in such a highly transcribed, gene-rich region, the closely spaced MARs provide a degree of structural stability or simply serve to keep the genes appropriately organized. They may also be points at which topoII may act to initiate decondensation or condensation of the loops, as a means of initiating or shutting down transcription.

Chromosome bands and functional domains

If metaphase chromosomes are stained with commonly used DNA-binding dyes, they reveal little in the way of structural detail. However, if they are treated in specific (and still rather mysterious) ways prior to staining, then regions of differing staining intensity are seen along the chromosome arms. One commonly used method involves fixing metaphase cells in 3:1 methanol:acetic acid, dropping them onto glass slides and treating for a few minutes with a proteolytic enzyme (trypsin is the one usually used) before staining with Giemsa. This procedure generates chromosomes with a striking pattern of light and dark bands. The latter are designated G-bands (Fig. 5.15). If chromosomes on slides are subjected, not to trypsin, but to hot alkali prior to staining with Giemsa, strongly staining bands, known as C-bands, are seen at and around the centromeres of all human chromosomes. They represent regions of centric heterochromatin. The properties of heterochromatin are discussed at length in Chapter 9 but for now it should be noted that C-bands are enriched in repetitive sequence elements, replicate late in S-phase and contain very little in the way of coding DNA.

The pattern of G-bands is exactly the same on the two sister chromatids of each chromosome. More importantly, within the same species, the G-banding pattern is reproducible from one individual to another and also from one cell

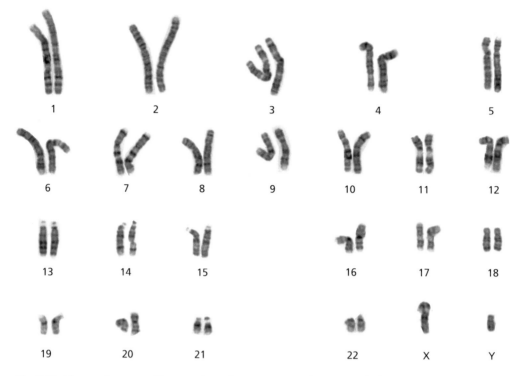

1   2   3   4   5

6   7   8   9   10   11   12

13   14   15   16   17   18

19   20   21   22   X   Y

Fig. 5.15 A human karyotype. The metaphase chromosomes used to prepare the karyotype were obtained from peripheral blood lymphocytes of a human male and were trypsin-digested and stained with Giemsa to reveal G-bands. In this preparation, the sister chromatids (the products of DNA replication in S-phase) have not separated and lie side-by-side. The maternal and paternal homologues of each chromosome share the same banding pattern, but each chromosome type is unique. (Karyotype prepared by Malcolm Taylor.)

type to another. This is despite the fact that the degree of compaction of metaphase chromosomes varies widely both between different preparations and from one spread to another on the same slide. Metaphase chromosomes are large and easily distorted. Although longer chromosomes tend to reveal more bands than more compact ones, these can usually be attributed to the separation of single large bands into several smaller bands as the chromosome extends. The reproducibility of the G-banding pattern suggests strongly that, even though the bands themselves are revealed only following treatment *in vitro*, they must reflect some underlying structural feature of metaphase chromosomes.

Evidence to support this came from the finding that banding patterns resembling G-banding could be obtained by very different pre-treatment regimes. Protocols could be devised that resulted in relatively strong staining (with Giemsa) specifically of the regions in between G-bands, now known as R-bands (because they are revealed by a Reverse staining procedure). Thus, the *distribu-*

tion of bands is not influenced by the pre-treatment, only their relative staining intensities. A further piece of evidence comes from the observation that the distribution of G-bands along metaphase chromosomes resembles the distribution of regions of apparently more condensed chromatin along chromosomes at the pachytene stage of meiosis. These regions, known as chromomeres, are visible by light microscopy in native, pachytene chromosomes. The fact that the same (or very similar) bands are present in both mitotic and meiotic chromosomes suggests that they represent a fundamental structural feature that, although it may be enhanced in mitotic chromosomes by various banding techniques, is not generated *de novo* by such techniques. Which raises the question of what this structural feature might be.

There is evidence to suggest that the base composition of chromosomal DNA may play a role in chromosome banding. Staining chromosomes with the DNA-binding fluorochrome quinacrine, without any pre-treatment other than the usual fixation, gives a pattern of brightly staining bands (called Q-bands) that correspond to G-bands. The explanation for this is that the fluorescence intensity of quinacrine is reduced in proportion to the frequency of G–C base pairs, i.e. GC-rich regions stain weakly because GC base pairs absorb, or 'quench', the emitted fluorescent light, whereas AT-rich regions stain relatively brightly. Detailed studies with quinacrine and other DNA-binding fluorochromes showed that G-bands are consistently richer in AT base pairs than R-bands. However, it seems unlikely that base composition *alone* can account for the banding phenomenon. The difference in base composition between G-bands and R-bands is too small to account, by itself, for the striking differences in fluorescence intensity shown by quinacrine and some other fluorochromes. A likely explanation is that the difference in base composition between R-bands and G-bands results not from the small *overall* difference, but instead from the presence of subregions that are very GC-rich or very AT-rich. Bright or dim labelling of these subregions could account for the overall differences in fluorescence intensity. As discussed below, there is now strong evidence that such GC-rich and AT-rich sequence elements are indeed present in R-bands and G-bands. R-bands have been shown to contain relatively high levels of a family of GC-rich repetitive DNA sequence elements known as SINES, whereas G-bands contain relatively high levels of an AT-rich family known as LINES. (We come across these elements again in Chapter 9.) In addition, the fact that G-banding can be induced by treatment with proteolytic enzymes suggests that proteins are somehow involved. As the great majority of proteins (including all or most of the histones) are removed from metaphase chromosomes by fixation in methanol:acetic acid, the protein(s) involved are likely to be tightly associated with chromosomal DNA.

Loops, SARs/MARs and chromosome bands

A model has been proposed that uses the concepts of DNA looping and scaf-

fold/matrix-associated DNA to explain the properties of metaphase chromosome bands. (Those who work with metaphase chromosomes tend to refer to
DNA sequences that attach to the chromosome scaffold as SARs, but they are
the same family of sequences as the matrix attachment regions, MARs, dealt
with earlier.) The model is based primarily on sophisticated microscopical
analysis of chromosomes labelled either with daunomycin, a fluorochrome that
preferentially labels very AT-rich DNA, or with a mixture of the general DNA-
binding fluorochrome YOYO and the nonfluorescent dye, Methyl Green. The
latter binds preferentially to AT-rich DNA and quenches the fluorescence of
YOYO, making the mixture (YOYO–MG) highly specific for GC-rich DNA.
When muntjac chromosomes (which are more extended than those of other
mammals) were stained with these fluorochromes, daunomycin was found to
label G/Q-bands strongly and R-bands weakly, as expected. However, the staining also revealed a coiled structure within the G-bands (Fig. 5.16). YOYO–MG
labelled the R-bands preferentially. The coiled structure within G-bands can be
interpreted as a chromosome scaffold closely associated with very AT-rich
(SAR-rich) DNA. The high frequency of SARs gives relatively small loops. In
contrast, in the R-band regions the SARs occur less frequently and the relatively
GC-rich loops are bigger (Fig. 5.16). The model does not require any great difference in DNA amount or compaction between R-bands and G-bands, just
packaging into loops of different sizes and a different scaffold arrangement
relative to the chromosome axis. The proposed association between the AT-rich
DNA regions and the chromosome scaffold was supported by showing that

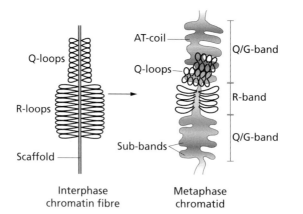

Fig. 5.16 A model for metaphase chromosome structure based on chromatin loops of
differing size and base composition. AT-rich DNA regions are associated with the
chromosome scaffold to form loops of differing sizes. The scaffold with its associated DNA
coils up in metaphase ('AT coil'). The smaller Q-loops form the more condensed chromatin of
Q/G-bands, whereas R-loops form the more diffuse chromatin of R-bands. The chromatin of
R-bands and Q/G-bands is made up entirely of loops, only some of which are shown in the
diagram. (Redrawn from Saitoh and Laemmli, 1994.)

antibodies to the scaffold component topoII labelled the same AT-rich regions as daunomycin.

The functional significance of chromosome bands

The development of reproducible banding procedures for metaphase chromosomes represented a technical breakthrough of major importance for many different branches of genetics and medicine. It enabled individual chromosomes and chromosome regions to be identified by their own specific banding pattern, much as a bar code is used to identify products in supermarkets. Chromosome banding is used for diagnosis of congenital defects, for identification of chromosome translocations in tumour cells and in gene mapping, amongst many other things. For science and clinical medicine, it is a technique of immense value. However, intriguing questions remain about the functional significance of chromosome bands. For example, if they were simply to reflect the influence of DNA base composition on the DNA packaging machinery that condenses chromosomes through mitosis, then they could reasonably be labelled a secondary phenomenon of very limited interest in attempting to understand how cells work. Alternatively, if it were possible to show a correlation between bands and, for example, the location of sets of genes or gene families whose patterns of expression were somehow related, then they might be thought of as rather more interesting.

Studies of the distribution of genes along chromosomes have shown that, in general, more genes are located in R-bands than in G-bands. Current data suggest that about 80% of all genes are located in R-bands and 20% in G-bands, with housekeeping (i.e. ubiquitously expressed) genes being virtually absent from G-bands. The available data are increasing rapidly as genome mapping projects move ahead, but it seems unlikely that this estimate will change radically. However, it is worth bearing in mind that it is not easy to determine, with confidence, whether or not a gene is located in a G-band. The reason is that most genes are mapped by genetic linkage analysis and sequencing, while chromosome bands are located by microscopical analysis of metaphase chromosomes. It can be difficult to reconcile the different types of map. Even when a gene is located to a G-band microscopically by hybridization of a labelled DNA probe to banded chromosomes, it is still possible that analysis at higher resolution would split the G-band into sub-bands, leaving the gene now in a region between bands. It has been suggested that genes switch between G-bands and R-bands depending on their transcriptional activity, but there is no firm evidence for such a switch.

If, as seems likely, the difference between G-bands and R-bands is linked in some way to their base composition, then the preponderance of genes in R-bands may be an inevitable consequence of the fact that coding DNA and adjacent sequences are relatively GC-rich. All housekeeping (i.e. ubiquitously expressed) genes and 40% of tissue-specific genes are associated with GC-rich regions known as CpG islands (Chapter 10) and about 86% of all CpG islands are found in R-bands. Finally, there is a subset of R-bands, known as T-bands, that are both very GC-rich and very gene-rich. A higher density of SARs in relatively AT-rich, gene-poor regions than in relatively GC-rich, gene-rich regions, could lead to the type of looping pattern proposed in Fig. 5.16 as the basis for G/Q- and R-bands.

The chromatin composition of chromosome bands

It is experimentally difficult to compare the structure and composition of chromatin from G-band and R-band regions of the genome, but it is unlikely that there are fundamental differences in nucleosome structure between the two. However, one striking difference has been revealed by immunolabelling of metaphase chromosomes with antibodies to acetylated histones H3 and H4. These experiments have shown that both C-bands (i.e. centric heterochromatin) contain little or no acetylated histone. Regions along the chromosome arms are also strikingly underacetylated, giving a clear fluorescent banding pattern (Plate 5.1). There is some indication that the underacetylated regions correspond to G-bands, but the fact that very different chromosome preparation techniques must be used for G-banding and immunofluorescent labelling precludes a direct and detailed comparison of the two banding patterns. However, G-bands are relatively gene-poor, AT-rich and late replicating, all properties shown, to an even greater extent, by centric heterochromatin. In view of this, it would not be surprising if G-band chromatin was also underacetylated.

Bands in interphase chromosomes

While metaphase chromosome bands have provided useful insights into the structure of mammalian genomes, they suffer from the disadvantage that metaphase is, for those who study gene expression, not the most interesting phase of the cell cycle. Transcription is minimal and chromosomes are packaged so as to complete a very specific task, namely division of the genome into daughter cells. Could it be that chromosome bands are a property of mitotic chromatin with no relevance to chromatin function in interphase cells?

Although microscopical techniques are of little value in examining the structure of chromatin in most interphase cells, there are a few special situations in which this can be done. These arise in cells in which several rounds of DNA replication occur without intervening cell divisions. This gives rise to giant, polytene chromosomes in which a thousand or more DNA strands are exactly aligned, side-by-side in a single chromosome. The most common model systems for studying these chromosomes are salivary glands from larvae of the fruitfly *Drosophila* or the midge *Chironomus*. These chromosomes are a joy to work with in that, once you have mastered the technique of transferring them to slides, they are easily seen down the light microscope and, when stained with more or less any DNA stain, or even when viewed by phase contrast, they show a reproducible pattern of light and dark bands (Fig. 5.17). There are about 5000 such bands in *D. melanogaster* and very detailed maps have been available for many years. These maps can be used to define the location of specific genes and chromosome domains. It is important to note that these are interphase chromosomes with the usual complement of transcriptionally active genes. Genes that are rapidly transcribed, such as certain genes inducible by hormones or heat shock, can be seen microscopically as expanded regions known as puffs (Fig. 5.17) and, as discussed in the next chapter, continue to provide a unique model system for the analysis of transcription by microscopy.

So, interphase chromatin (at least in polytene cells) does exhibit a banded pattern. But what do the bands represent, and do they share any properties with the bands seen in mammalian metaphase chromosomes? Perhaps the first point to note is that the chromatin in the bands of polytene chromosomes in insects is much more compacted than that in the interbands, probably by about 50:1. This is readily seen in scanning electron micrographs of polytene chromosomes (Fig. 5.18). It is also interesting to note that levels of H4 acetylation are often higher in polytene interbands than in bands. Regions that stain intensely with Hoechst often label only weakly with antibodies to acetylated H4 *and vice versa.* (Compare the distribution of labelling along chromosome arms in Fig. 5.17a and b.)

Three lines of evidence show that polytene chromosome bands are stable structures. Firstly, where it has been possible to prepare maps from polytene chromosomes from different tissues, only minor variations have been observed from one tissue to another. Secondly, banding patterns have been highly conserved through evolution and, thirdly, they survive chromosome translocation. These properties suggest that the differences between bands and interbands lie at the DNA level. However, there is no evidence for differences in base composition between band and interband DNA nor, as yet, have sequence comparisons revealed (inter)band-specific motifs. It is also possible that such will be found now that the *Drosophila* genome sequencing project is complete. It is possible that differences are due to structural motifs, such as cruciforms or triple helices, each of which may be defined by a variety of DNA sequence elements and whose identification will require sophisticated analytical techniques.

Nuclear domains and structure in the interphase nucleus

With the exception of the special, polytene cells noted above, the analysis of structure–function relationships in the interphase nucleus was, for many years, beset with intractable experimental problems resulting, almost exclusively, from the lack of the necessary, specific molecular probes. Three experimental approaches have transformed the study of the interphase nucleus. The first is

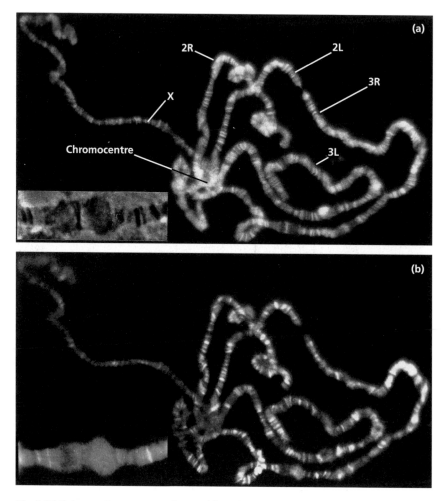

Fig. 5.17 Polytene chromosomes of *Drosophila melanogaster*. (a) Chromosomes stained with the DNA-binding fluorochrome Hoechst 33342. For chromosomes 2 and 3, the chromosome regions either side of the centromere (which is close to the middle) are distinguished as L and R. The X chromosome centromere is at one end. Note the intense staining of the chromocentre (lower left), the region where the heterochromatin-rich centromeres cluster together. The small chromosome 4 associates closely with the chromocentre and is not easily distinguishable in this photograph. Chromosome bands are intensely stained and interbands are weakly stained, reflecting their relative DNA content. The larvae were subjected to a brief heat shock (37°C) before chromosome preparation and the *HSP70* heat shock genes at positions 87A and 87C are transcriptionally active, forming two prominent puffs (bottom centre). The expanded DNA of the puffs stains weakly with Hoechst. An enlargement of the 87A–C region viewed under phase contrast is inset. The puffed regions are phase-dense due a high concentration of RNA transcripts and associated proteins. (b) The same chromosome spread as above labelled by indirect immunofluorescence with a rabbit antiserum to H4 acetylated at lysine 8. Note the generally weak labelling of the chromocentre and the lack of correspondence between the Hoechst and antibody labelling patterns. The distribution of acetylated H4 does *not* reflect that of DNA (chromatin). The larger heat shock puff (87C), containing three copies of the *HSP70* gene, consistently labels more strongly than the smaller (87A) with two copies (inset). The reasons for this are not clear and interpretation is complicated by the accumulation of RNA and protein at the puffs, with possible effects on antibody accessibility. (Immunolabelling by Richard Simpson.)

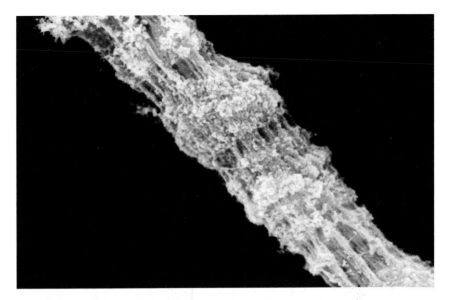

Fig. 5.18 Scanning electron micrograph of a polytene chromosome. The condensed chromatin of the bands and the more extended chromatin of the interbands is clearly distinguishable. Magnified about 4200×. (Preparation by Klaus Pelling, photograph supplied by Terry Allen.)

the use of labelled DNA and RNA probes, together with the appropriate *in situ* hybridization techniques, to detect specific chromosomes, genes, DNA elements and RNA transcripts within the nucleus. The second is the use of antibodies to identify specific nuclear components by immunolabelling, and the third is the use of nonradioactive DNA and RNA precursors modified in such a way that they can be detected by specific antibodies, usually labelled with fluorescent dyes. The different labelling methods can often be used together on the same nucleus with different coloured fluorochromes being used to provide distinguishable signals. When combined with sophisticated microcopes and image analysis systems, these approaches can provide a three-dimensional picture of nuclear organization and function.

Examination of a conventionally fixed and stained interphase nucleus by light microscopy will show one or more large densely staining bodies, sometimes with a granular appearance. These are the nucleoli, the sites of synthesis of ribosomal RNA (rRNA) and ribosome assembly. In some cells you will also see dark staining patches adjacent to the nuclear envelope. These are regions of heterochromatin, the highly condensed, gene-poor DNA that we encountered earlier as C-bands on metaphase chromosomes. These two very simple observations tell us at once that the genome is not randomly distributed through the nucleus and that certain genes (i.e. the rRNA genes on chromosomes 13, 14 and 15 in humans) and DNA elements (centric heterochromatin) can take up specific positions. This argument has been enormously strengthened recently by the use of chromosome 'paints' (DNA probes that recognize only a single chromosome) to show that individual chromosomes occupy relatively defined and nonoverlapping territories in the interphase nucleus. Furthermore, by using nonradioactive RNA precursors to allow microscopical detection of RNA transcripts, evidence has been assembled to indicate that transcription tends to occur at the surface of chromosome domains. This is an attractive model in that the interchromosomal spaces may provide channels through which RNA transcripts are transported to the nuclear pores and thence to the cytoplasm. The transport and processing of RNA transcripts is a major task for the cell nucleus and one whose size and complexity matches that of transcriptional control. It is interesting to note that some of the nuclear structures described some years ago by electron microscopists have recently been shown to be involved in RNA processing, particularly splicing. These include such mysterious elements as coiled bodies and interchromatin granule clusters. Fortunately, consideration of RNA processing reactions falls outside the scope of this book, but it is important to note that these structures provide another very good example of the importance of higher-order organization within the interphase nucleus. A simple model outlining current concepts of nuclear organization is shown in Fig. 5.19. The organization and distribution of chromatin domains in the nucleus is central to many aspects of gene regulation and will be revisited in later chapters.

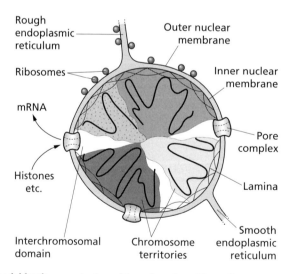

Fig. 5.19 A model for the organization of the cell nucleus. Heavy lines represent the paths of chromosome axes. Chromatin loops extending from the axis (scaffold?) fill the shaded regions, forming chromosome territories. There is evidence that active genes are located on the outer margins of chromosome territories. RNA transcripts are processed in specific regions within the interchromosome domains and transported to the cytoplasm via nuclear pore complexes. (Based on the model proposed by L. Gerace, 1984.)

Further Reading

Higher-order chromatin structure

Belmont, A. S. & Bruce, K. (1995) Vizualization of G1 chromosomes—a folded, twisted, supercoiled chromonema model of interphase chromatid structure. *J. Cell Biol.*, **127:** 287–302.

Pederson, D. S., Thoma, F. & Simpson, R. T. (1986) Core particle, fibre and transcriptionally active chromatin structure. *Ann. Rev. Cell Biol.*, **2:** 117–147.

Williams, S. P., Athey, B. D., Muglia, L. J., Schappe, R. S., Gough, A. H. & Langmore, J. P. (1986) Chromatin fibres are left-handed double helices with diameter and mass per unit length that depend on linker length. *Biophys. J.*, **49:** 233–248.

The nuclear matrix

Fey, E. G., Wan, K. M. & Penman, S. (1984) Epithelial cytoskeletal framework and nuclear matrix–intermediate filament scaffold: three dimensional organization and protein composition. *J. Cell Biol.*, **98:** 1973–1984.

Van Driel, R., Wansink, D. G., van Steensel, B. *et al.* (1995) Nuclear domains and the nuclear matrix. *Int. Rev. Cytol.*, **162A:** 151–189.

DNA loops and matrix attachment

van Drunen, C. M., Sewalt, R. G. A. B., Costerling, R. W. *et al.* (1999) A bipartite sequence element associated with matrix/scaffold attachment regions. *Nucl. Acids Res.*, **27:** 2924–2930.

Mirkovitch, J., Mirault, M.-E. & Laemmli, U. K. (1984) Organization of the higher-order chromatin loop: specific DNA attachment sites on the nuclear scaffold. *Cell*, **39:** 223–232.

Mullinger, A. M. & Johnson, R. T. (1980) Packing DNA into chromosomes. *J. Cell Sci.*, **46:** 61–86.

Paulson, J. R. & Laemmli, U. K. (1977) The structure of histone-depleted metaphase chromosomes. *Cell*, **12:** 817–828.

Pienta, K. J. & Coffey, D. S. (1984) A structural analysis of the role of the nuclear matrix and DNA loops in the organization of the nucleus and chromosome. *J. Cell Sci.*, Suppl. 1, 123–135.

Chromosome bands

Craig, J. M. & Bickmore, W. A. (1993) Chromosome bands—flavours to savour. *BioEssays*, **15:** 349–354.

Korenberg, J. R. & Rykowski, M. C. (1988) Human genome organization: Alu, Lines, and the molecular structure of metaphase chromosome bands. *Cell*, **53:** 391–400.

Saitoh, Y. & Laemmli, U. K. (1994) Metaphase chromosome structure: bands arise from a differential folding path of the highly AT-rich scaffold. *Cell*, **76:** 609–622.

Chromosome compaction

Gasser, S. M. (1995) Coiling up chromosomes. *Curr. Biol.*, **5:** 357–360.

Nuclear domains and organization

Jackson, D. A. & Cook, P. R. (1995) The structural basis of nuclear function. *Int. Rev. Cytol.*, **162A:** 125–149.

Kurz, A., Lampel, S., Nickolenko, J. E. *et al.* (1996) Active and inactive genes localize preferentially in the periphery of chromosome territories. *J. Cell Biol.*, **135:** 1195–1205.

Lamond, A. I. & Earnshaw, W. C. (1998) Structure and function in the nucleus. *Science*, **280:** 547–553.

Spector, D. L. (1992) Macromolecular domains within the cell nucleus. *Ann. Rev. Cell Biol.*, **9:** 265–315.

Chapter 6: Transcription in a Chromatin Environment

Introduction

In all eukaryotic cells, chromatin is necessary for the packaging of DNA into the cell nucleus and for its efficient and accurate replication and segregation into daughter cells during each cell cycle. These functions alone are enough to make chromatin an essential component of eukaryotic life. From first principles, it seems inevitable that chromatin must also influence the other major function of the eukaryotic nucleus, namely gene expression. The intimate association of histones with DNA and the various levels of higher-order DNA packaging discussed in the previous chapter all influence the binding to DNA of transcription factors and other components of the transcription complex. (The way in which the nucleosome influences the binding of proteins to DNA is discussed in the next chapter.) Chromatin is something that the transcription apparatus must have learned to get along with from the very earliest stages of eukaryotic evolution.

These simple considerations bring us to a more difficult problem, namely, the need to establish the extent to which chromatin is an integral and necessary component of transcriptional control mechanisms. The significance of this problem may become clearer if two extreme views of the role of chromatin are considered. One holds that chromatin is primarily a DNA packaging device; it constitutes an obstacle that the transcription apparatus must overcome and its effects on gene expression, although unavoidable, are essentially *passive*. The other holds that chromatin is an integral component of mechanisms of transcriptional control and that this role has evolved in parallel with its packaging function; it plays an *active* role in control of gene expression. The functional difference between these two possibilities is similar to that between a pile of rocks and a set of traffic lights. Both can be equally effective, except in certain parts of Italy, at stopping traffic. But the traffic lights offer a level of flexibility and sophistication in regulating traffic flow that the pile of rocks cannot match. So, which category does chromatin fall into? The answer is probably both, because the two possibilities are not mutually exclusive. Specific effects of chromatin can result from either active or passive mechanisms, and piles of rocks do have some advantages. But the distinction is an important one. Unravelling a mechanism that involves chromatin as an active participant will lead to more useful insights into how genes are regulated than spending time on effects that are by-products of chromatin just being there. It is also worth mentioning that, in some cases, chromatin can exert a *positive* effect on gene expression. As discussed later, it can do this by folding DNA in such a way as to bring together protein binding sites and thereby facilitate useful protein–protein interactions.

This leads to a second important reminder, namely the need to bear in mind (without being intimidated or depressed by it) the complexity of the system we are trying to understand. A typical mammal has tens of thousands of genes in each of its cells, and each one of these genes must be regulated in a way appropriate to the needs of the cell in which it finds itself. This is not to say that there are likely to be tens of thousands of different mechanisms of gene regulation, but nor will there be just one. Even in a single-celled eukaryote such as yeast, some genes change their levels of activity through differentiation, the cell cycle or in response to environmental changes while others are expressed in all cells for most or all of the time. In multicellular organisms even more regulatory demands must be met. Chromatin is an essential element in the mechanisms used to address these demands, but the ways in which it is used will differ depending on the nature of the problem that has to be solved. It is particularly important to remember this when thinking about effects that are brought about by the nucleosome itself. In Chapter 3, I discussed the extraordinary conservation of this structure through evolution. The way in which the histone octamer packages DNA has not changed much over hundreds of millions of years. However, in terms of its role in regulating gene expression, there is no reason why this highly conserved structure should not be used in radically different ways to deal with the regulatory requirements of different genes.

In what follows, there are examples that introduce possible mechanisms of gene regulation by chromatin. They have been taken from a variety of model systems, ranging from yeast to fruitflies to mammals, each of which has its advantages and disadvantages in studying a particular aspect of the problem. In all cases our understanding of the molecular mechanisms involved is incomplete. It may turn out that, in some cases, different organisms use chromatin in quite different ways. But it is more likely that, in most cases, a mechanism used by one organism will also be used by another, although possibly in a different context or for a different biological purpose. Evolution makes use of what comes to hand and the range of mechanisms found can be looked upon as a sort of molecular tool kit that organisms can reach into for carrying out particular tasks.

Genes are packaged into nucleosomes, even when they are being transcribed

One superficially simple way of circumventing any possible chromatin-induced complications for the transcription apparatus would be to avoid packaging genes as chromatin. Coding DNA, after all, constitutes only a small proportion, perhaps 1–2%, of total DNA in mammals and, as we will see later, mechanisms do exist for the differential packaging of different regions of the genome and even for keeping some of them nucleosome-free. However, it is clear that genes themselves are packaged within the nucleus as chromatin. We know this because if interphase nuclei are gently digested with micrococcal nuclease, then both coding DNA (genes) and noncoding DNA are fragmented into oligonucleosome-sized pieces, i.e. both types of DNA are wrapped up in nucleosomes. This is illustrated in Fig. 6.1. They can also both be isolated as 11 S particles by centrifugation through sucrose gradients.

However, significant changes in chromatin structure do occur as genes switch from quiescent to a transcriptionally active state. For very rapidly transcribed genes this change is detectable by simple micrococcal nuclease digestion, with the inactive gene giving a sharp, clearly defined ladder of DNA bands and the active gene a much more diffuse pattern of bands. An example is shown in Fig. 6.1. Two immediate conclusions can be drawn from these differences in nuclease digestion pattern. The first is that transcription disrupts the spacing of nucleosomes and possibly their association with DNA, giving rise to the smeared pattern of DNA bands. This is not unexpected given what we know

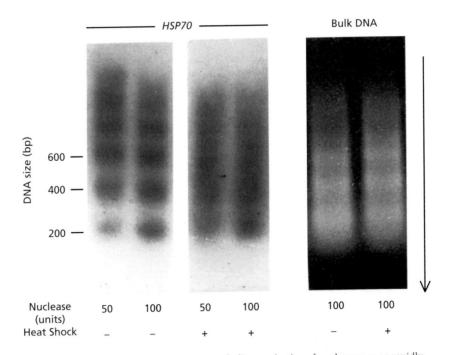

Fig. 6.1 Micrococcal nuclease digestion reveals disorganization of nucleosomes on rapidly transcribed genes. If *Drosophila* cells grown in tissue culture are briefly exposed to a temperature of 30°C or more, high levels of transcription are initiated at a specific set of genes, the heat shock genes. Micrococcal nuclease digestion of nuclei from heat shocked and control cells reveals no difference in the digestion pattern of bulk chromatin (ethidium-stained gel, R-hand side). Southern blotting of DNA from the same samples with a probe to the HSP70 heat shock gene shows that the non-transcribed gene produces a clear ladder of bands typical of evenly spaced nucleosomes (L-hand side). In contrast, the active gene (i.e. after heat shock) shows a diffuse labelling pattern, indicative of changes in the spacing, and possibly the structure, of nucleosomes along the gene. (Data from Rebecca Munks.)

about the mechanism of action of RNA polymerases. Even a relatively simple, prokaryotic polymerase, during transcriptional elongation, is closely associated with about 50 bp of DNA. This DNA is deeply buried within the enzyme, and 10–20 bp at the active site has been transiently unwound (i.e. the two strands have been separated). In view of this, it is inconceivable that the polymerase should simply progress around the nucleosome, causing neither displacement nor structural rearrangements. The strategies that polymerases might use to negotiate their way around nucleosomes are discussed later. Smearing of DNA bands following nuclease digestion tends to be seen only in very actively transcribed genes, such as the ribosomal RNA and heat shock genes, with a high density of polymerases. This suggests that the structural changes in the nucleosome induced by polymerase are transient and reversible and that the nucleosomes along genes with a low polymerase density are mostly normal.

The second conclusion from results of the type shown in Fig. 6.1 is that nucleosomes, or nucleosome-like structures, can remain associated with the coding DNA during transcription. If they did not, then the DNA of the gene would be very rapidly digested and such fragments as did survive would not show a residual ladder of DNA bands. This important conclusion has been confirmed by experiments showing that core histones can be cross-linked to transcribing DNA by treatment with formaldehyde. Visual confirmation has come from electron microscopical (EM) studies of ribosomal RNA genes rapidly transcribed by Pol I. These genes often reveal nucleosome-sized particles in between RNA polymerase molecules. The most convincing studies use pre-treatment with psoralen, a detergent that introduces cross-links into the linker DNA, but not DNA of the nucleosome core particle. If the DNA is deproteinized and denatured prior to EM analysis, sites that originally contained nucleosomes are revealed as 'bubbles' of single-stranded, noncross-linked DNA (Fig. 6.2). This approach gets around the problem that, in material prepared for electron microscopy, nucleosomes and Pol I are similar in size and rather hard to distinguish.

The structure of chromatin along a transcribing gene

All methods for looking at the structure of chromatin before, during and after transcription are indirect. We can, as discussed later in this chapter, measure the sensitivity of defined regions to nuclease digestion or chemical cleavage and we

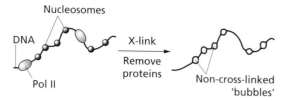

Fig. 6.2 Psoralen cross-linking reveals the positions of nucleosomes in transcribing DNA.

can use chromatin immunoprecipitation (ChIP) to decide whether or not histones are present and what their state of modification is. But none of these procedures can relate the structure of chromatin *in situ* to the structures discussed in Chapters 3 and 4. Are the 10 nm, beads-on-a-string fibres or 30 nm fibres, identified by EM of spread chromatin, present during transcription? Questions such as this have, as yet, no clear answers, largely because we have no reliable methods for visualizing structures of the size and complexity of chromatin fibres within such a densely packed structure as the interphase nucleus. The spreading and fixation required for EM visualization usually preclude analysis of local chromatin structures. However, some important insights have come from studies of a useful model system, namely the rapidly transcribed Balbiani Ring genes of the midge *Chironomus*. The salivary glands of *Chironomus* larvae have polytene chromosomes that are even larger than those in *Drosophila* and that give an equally striking and reproducible banding pattern. The smallest chromosome has two very prominent puffs, known as Balbiani Rings, that are the sites of the rapidly transcribed genes encoding secretory proteins that *Chironomus* larvae use to spin a protective tube. Figure 6.3 shows an EM picture of *Chironomus* chromosome IV with the Balbiani Rings 1, 2 and 3. Functional transcription units and new transcripts packaged with proteins for transport and processing can be seen at higher magnification in Fig. 6.4. Bertil Daneholt and colleagues have used electron microscopy to investigate the structure of transcribing chromatin in these puffs. They conclude that the rapidly transcribing chromatin exists as an extended 5 nm fibre (Fig 6.5A,B). However, where the gap between polymerases is more extended, nucleosomes, 10 nm fibres (Fig. 6.5C)and even 30 nm fibres (Fig 6.5D) can re-form. By immuno-staining with antibodies coupled to gold particles (detectable by EM), it has been shown that histone H1 remains closely associated with rapidly transcribed chromatin. The message from these results is that transcribing chromatin is structurally

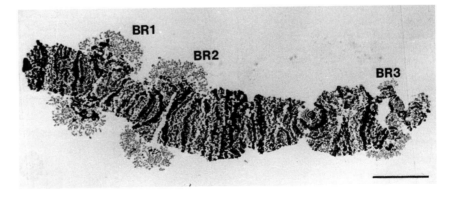

Fig. 6.3 Electron micrograph of an isolated chromosome IV from salivary glands of midge *Chironomus tentans*. The three giant puffs, Balbiani Rings (BR) 1, 2 and 3 are indicated. Bar= 5 μm. (From Björkroth *et al.* 1988.)

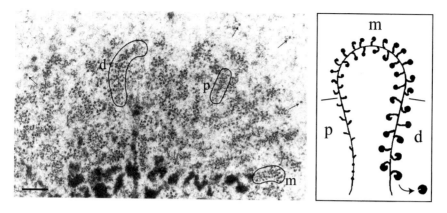

Fig. 6.4 RNA transcripts from Balbiani Ring genes are packaged with protein even before transcription is complete. Partially packaged transcripts can be seen in electron micrographs of Balbiani Ring genes as electron-dense spots. Protein–RNA complexes become larger as transcription proceeds, and can be designated as proximal (p), medial (m) or distal (d) from the transcription start site, as shown in the diagram. Examples of each stage are circled in the micrograph. Bar = 1 μm. (From Skogland *et al.*, 1983. *Cell* **34**: 847–855.)

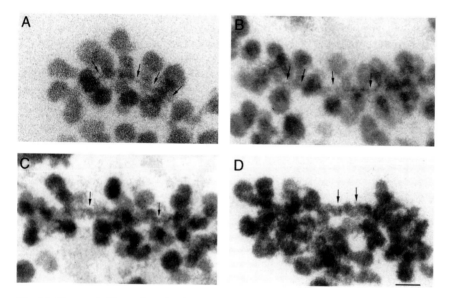

Fig. 6.5 Chromatin fibres of varying diameters can be visualized in actively transcribing Balbiani Ring genes. The four electron micrographs show partially complete RNA transcripts complexed with protein (dense spots). Fibres with diameters of 5 nm (A and B), 10 nm (C) and 30 nm (D) can sometimes be detected between the transcripts (arrows). The composition of such fibres cannot be determined from electron micrographs alone, but they may represent naked DNA (5 nm) or DNA packaged at different levels of chromatin structure (10 nm and 30 nm). Such data suggest that DNA can be assembled into chromatin between transcription units and that ongoing chromatin assembly and disassembly is an integral part of the transcription process. Bar = 50 nm. (From Björkroth *et al.* 1988.)

dynamic and that regions lacking polymerase can rapidly re-form higher-order structures. As the great majority of genes are transcribed much more slowly than the Balbiani Ring genes (i.e. have a much lower density of polymerases), it may be that chromatin fibres of 30 nm or above are the norm, even when transcription is under way.

To summarize, an accumulation of experimental data shows that nucleosomes and polymerases *can* coexist along a transcribing gene. But it is important to emphasize that they do not *always* do so (and to restate the importance of avoiding generalizations). Along the most actively transcribed genes in yeast, polymerases can occur every 100 bp or so, leaving no room for nucleosomes, although not ruling out the continued presence of histones in one form or another. It is also important to emphasize that, although nucleosomes can be retained on genes and on transcribing DNA, they are not 'transparent' to the transcription machinery, as has been claimed in the past. This was elegantly demonstrated by the genetic experiments outlined in the next section.

Genetic experiments in yeast show the importance of histones for gene regulation

A conclusive demonstration that histones play an important role in gene regulation *in vivo* came from the construction of histone-deficient mutants in the yeast *Saccharomyces cerevisiae*. Amongst its many other advantages, *S. cerevisiae* has only two copies of the genes for each of the core histones, in contrast to the hundred or more copies found in higher eukaryotes. This means that they are amenable to mutational analysis. In the late 1980s, experiments were initiated in the labs of Michael Grunstein and Fred Winston to explore the effects of histone depletion on gene expression in yeast. The question was a deceptively simple one: can patterns of gene expression be altered by reducing the amount of core histones synthesized by a cell and, hence, the density of nucleosomes along the DNA? If the nucleosome did play a role in regulating gene expression then the answer would be yes, if it did not, i.e. if the nucleosome was simply ignored by the transcriptional machinery, then the answer would be no.

To perform this experiment, yeast strains were constructed in which the endogenous H4 genes had been deleted and in which H4 was supplied by a stable, extrachromosomal plasmid where the H4 gene was under the control of the GAL1 promoter. When cells were grown in medium containing galactose, then the plasmid-born H4 gene was *on* and the cells had normal levels of H4. However, when cells were grown in glucose instead of galactose, then transcription of the H4 gene was reduced and levels of H4 fell. The experimental system is outlined in Fig. 6.6.

A characteristic of yeast that is well known to those who work with it is its remarkable resilience. It will tolerate insults that more complex eukaryotes will not. In some circumstances this can be a nuisance, but in the present experiments the organism's resilience enabled it to continue to grow and divide, albeit

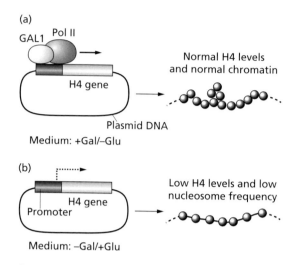

Fig. 6.6 A yeast plasmid containing the *H4* gene under the control of a GAL1 promoter is active in the presence of galactose and inactive in its absence. Mutant cells in which the two endogenous *H4* genes have been disabled can be made dependent on the plasmid for their supply of H4. Removal of galactose from the growth medium will result in a dramatic reduction in H4 synthesis and progressive reduction in nucleosome frequency, as chromatin packaging is unable to keep pace with DNA replication.

rather more slowly than normal, even in the presence of significant reductions in the level of H4. This property means we can be reasonably confident that any effects observed are attributable to histone depletion and do not simply reflect the aberrant behaviour of sick and dying cells. Also, chromatin in the histone-depleted cells was found to show increased sensitivity to nuclease digestion, indicating that the reduction in histone synthesis did, as expected, lead to an overall reduction in nucleosome density.

The effects of H4 depletion on gene expression were unequivocal. Histone depletion could cause de-repression of endogenous genes. Genes whose expression was up-regulated were all inducible and included *GAL1* (induced by galactose), *HIS3* (induced by amino acid deprivation) and *PHO5* (a gene induced by low levels of inorganic phosphate and about which much will be said later). Some inducible genes, such as *CUP1*, were affected to a lesser extent than these, and *constitutive* genes, i.e. genes that are switched on all the time, were unaffected by nucleosome depletion. But despite these not unexpected complexities, the take-home message was remarkably clear: histone depletion de-repressed certain important genes and the nucleosome must be an integral component of at least some gene regulatory mechanisms.

Changes in chromatin structure precede gene activation

Over a period of more than 20 years, endonucleases have been used to probe the structure of transcriptionally active and inactive chromatin. The value of nucleases derives from the fact that their ability to cut DNA is exquisitely sensitive to the way in which it is packaged as chromatin. Cutting generally is inhibited by chromatin, either through shielding by the nucleosome itself or through folding into higher-order structures and steric exclusion. Alternatively, chromatin-induced folding can introduce bends at specific sites in DNA that may, in some cases, be more susceptible to cutting. It is important to note that the two most commonly used endonucleases, DNAseI and micrococcal nuclease, differ in a number of important properties. Most importantly, whereas DNAseI can make single-stranded cuts (nicks) in nucleosomal DNA, micrococcal nuclease can only cleave linker DNA (Fig. 3.4). Like other enzymes, their activity is also influenced by the experimental conditions under which they are used. These considerations make it easier to understand why a specific gene or genomic region can sometimes be more sensitive to one enzyme than to the other.

Experiments on the globin and ovalbumin genes in chick erythrocytes were the first to show that active and inactive genes differed in their susceptibility to digestion by DNAseI, although not always to digestion by micrococcal nuclease. The work was extended to show that in oviduct cells, where the ovalbumin gene was active, it became more DNAseI sensitive. Subsequent work has identified similar differences in nuclease sensitivity in many different genes and has confirmed that an increase in nuclease sensitivity, usually of about 10-fold, is a general property of transcriptionally active chromatin (Fig. 6.7). Developmental studies showed that the chromatin of specific genes switched from nuclease resistance to nuclease sensitivity as development progressed and, significantly, that these changes could precede the developmental stage at which up-regulation of transcription occurred. For example, the genes encoding the immunoglobulin heavy and light chains undergo major changes in transcriptional activity (along with some unusual DNA rearrangements) during the maturation of antibody-producing lymphocytes. Detailed studies of the mouse κ light chain gene showed that DNAseI sensitivity preceded up-regulation of transcription and was related to transcriptional *competence* rather than transcriptional activity *per se*. It is generally now accepted that changes in nuclease sensitivity are a sign that the chromatin structure of the gene has shifted to one that will allow transcription to begin when conditions are right. That nuclease sensitivity can occur independently of transcription is supported by several observations showing that nuclease-sensitive domains often extend far beyond the gene itself. For example, the chicken ovalbumin gene is embedded in a DNAseI-sensitive domain extending over 80 kbp, while the gene itself, including known promoter and control regions, is less than 12 kb long. Chromatin changes are not merely a consequence of transcription, but occur independently and prior to its initiation.

The structural alterations that underlie changes in nuclease sensitivity are likely to be complex and are not well understood. However, modifications to individual chromatin components are now being studied at the molecular

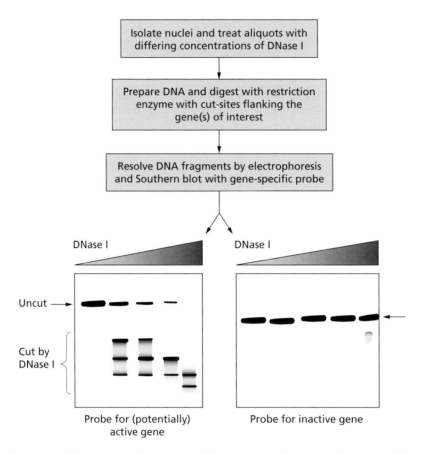

Fig. 6.7 Sensitivity to DNaseI digestion provides a measure of the transcriptional potential of chromatin.

level and correlated with changes in nuclease sensitivity. These include histone modifications, H1 stoichiometry, association of nonhistone proteins and DNA methylation. A particular level of nuclease sensitivity is likely to be the end result of a collection of coordinated and interacting chromatin modifications.

Increased histone acetylation can precede or accompany the onset of transcription

Almost 40 years ago, it was shown that when lymphocytes were stimulated in culture to move them from quiescence to a transcriptionally active state, there was an accompanying increase in the overall level of histone acetylation. Over succeeding years, various experimental approaches have been used to confirm this general association, and to try and define the ways in which it might be operating. Is acetylation a cause or a consequence of the chromatin changes that accompany the onset of transcription? If the former, what are the mechanisms by which it operates? Much of the early work involved the fractionation of transcriptionally active chromatin by biochemical means. In general, this confirmed the original conjecture that transcriptionally active chromatin was more highly acetylated than bulk chromatin, but was never able to exclude the possibility that acetylated and active chromatin simply co-purify, rather than being one and the same. A breakthrough came with the development of experimental approaches based on the use of antibodies capable of distinguishing acetylated and nonacetylated histones. The approach is described in Box 2. Using developing chicken erythrocytes as a model system, Hebbes and colleagues in 1988 showed that the acetylated fraction contained DNA from the globin genes, active in this cell lineage, but not from the inactive ovalbumin gene. Subsequent studies showed that the region of increased acetylation extended across the entire β-globin domain, a region of more than 100 kb. The region of hyperacetylation corresponded closely with the domain of generalized DNaseI sensitivity across the same locus.

A crucial finding in these experiments was that this acetylated domain was present in erythrocyte precursors (but not other cell types), even at stages of development that preceded the onset of transcription of the β-globin gene. In other words, acetylation was defining the *potential* for transcription, rather than the existence of transcription itself.

More recent work using the chromatin immunoprecipitation (ChIP) approach has shown that, in addition to defining large chromatin domains, transient increases in acetylation can also accompany the onset of transcription itself, at least for some genes. These changes are very local, sometimes confined to one or two nucleosomes in the promoter region, and may involve specific histones. For example, depending on the gene, H3 may be acetylated to a greater extent than H4. These local changes in acetylation require the targeting of histone acetylating and deacetylating enzymes (HATs and HDACs, respectively) in the ways discussed in Chapter 8 and the pattern of acetylation put in place at any given promoter will depend on the specificities of the targeted enzymes. The molecular mechanisms by which histone acetylation facilitates or inhibits transcription remain something of a mystery. Acetylation across large chromatin domains may help maintain a more open (nuclease-sensitive) higher-order structure, possibly by inhibiting internucleosome cross-links of the type described in Chapter 4 or by inhibiting other tail–protein interactions. There is ample experimental evidence that high levels of acetylation alter higher-order chromatin structure *in vitro* and facilitate transcription factor binding to nucleosome DNA (Chapter 7).

Box 2 Chromatin immunoprecipitation (ChIP)

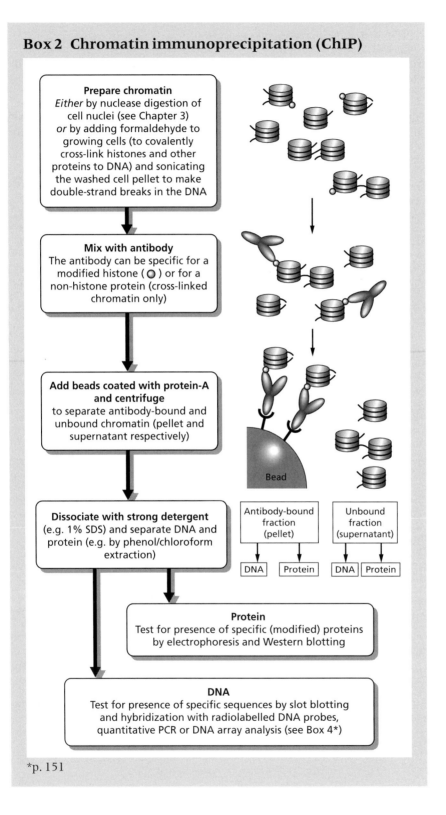

Prepare chromatin
Either by nuclease digestion of cell nuclei (see Chapter 3) *or* by adding formaldehyde to growing cells (to covalently cross-link histones and other proteins to DNA) and sonicating the washed cell pellet to make double-strand breaks in the DNA

Mix with antibody
The antibody can be specific for a modified histone (O) or for a non-histone protein (cross-linked chromatin only)

Add beads coated with protein-A and centrifuge
to separate antibody-bound and unbound chromatin (pellet and supernatant respectively)

Bead

Dissociate with strong detergent
(e.g. 1% SDS) and separate DNA and protein (e.g. by phenol/chloroform extraction)

Antibody-bound fraction (pellet)

Unbound fraction (supernatant)

DNA | Protein

DNA | Protein

Protein
Test for presence of specific (modified) proteins by electrophoresis and Western blotting

DNA
Test for presence of specific sequences by slot blotting and hybridization with radiolabelled DNA probes, quantitative PCR or DNA array analysis (see Box 4*)

*p. 151

DNaseI hypersensitive sites

By combining endonuclease digestion of chromatin or nuclei with a technique that combines nuclease digestion with a form of Southern blotting known as indirect end labelling (outlined in Box 3), it became possible to map variations in nuclease sensitivity (and hence changes in chromatin structure) across specific genes and genomic regions. This approach has identified precisely localized re-

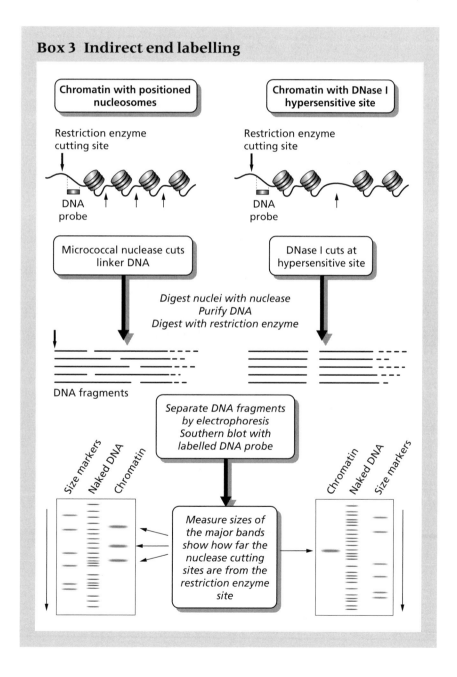

Box 3 Indirect end labelling

gions of the genome, usually a few hundred bp in size, that are extremely sensitive to digestion by both non-specific endonucleases and a variety of DNA-cleaving chemicals. The rate at which the DNA within such regions is broken down is at least an order of magnitude faster than that within the larger nuclease-sensitive regions discussed above. DNaseI digestion has been, and remains, the most commonly used method for detecting these regions and they are usually referred to as DNaseI hypersensitive sites (DHS).

The speed at which DHS are digested suggests that they are nucleosome-free and in some cases, where it has been possible to test this proposition experimentally, this has been proven correct. One of the most striking demonstrations has been through visualization of the SV40 virus isolated from mammalian cell nuclei. A virus may seem a rather unusual model for the study of eukaryotic chromatin, but it is one that has proven particularly useful. The 5243 bp circular DNA of the virus is packaged as chromatin and it has been shown, by nuclease digestion, to have a DHS located at a crucially important region (designated ORI) containing the replication origin, promoters and a binding site for the viral regulatory protein T-antigen. Examination of viral chromatin by electron microscopy shows a prominent, nucleosome-free region of about the right size (350 bp) and in the correct location. The absence of nucleosomes from DHS has been confirmed by an alternative approach in which cells are treated with formaldehyde, a reagent that efficiently cross-links histones and DNA. Cleavage of cross-linked chromatin and immunoprecipitation with antihistone antibodies failed to detect any histones associated with DHS within either the SV40 virus or the *Drosophila hsp70* gene. However, it is not safe to assume that every DHS will be nucleosome free, or even that this is the usual situation. We will see later how the assembly of multiple factors can generate a site of DNaseI hypersensitivity on DNA that is still associated with a nucleosome, albeit a rather unusual one. DHS are functionally (experimentally) defined and the simple existence of a DHS tells us little about the conformation or packaging of the DNA within it.

The functional significance of DHS

As more DHS were identified, it became apparent that they were often associated with DNA regions of known functional significance, such as promoters, enhancers and replication origins. This makes perfectly good sense in that these regions exert their functional effects through the action of sequence-specific binding proteins. What better way to ensure the efficient binding of such proteins than to remove the obstacle of the nucleosome? It is also clear that the DHS are used as an active method of regulation in that, while some are constitutive (i.e. are present all the time) others are present only at certain stages of development or in certain tissues and cell types. For example, the β-globin gene complex on human chromosome 11 contains both constitutive DHS and sites

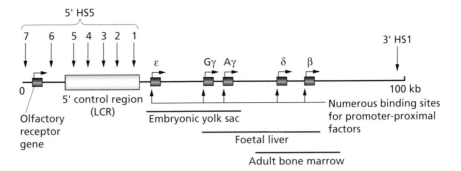

Fig. 6.8 The human β-globin gene cluster. Specific genes are expressed at defined times during foetal development, as indicated.

that are found only in erythroid cells in which the genes are actively expressed (Fig. 6.8).

At least some DHS are likely to represent regions of the genome that have been kept nucleosome-free in order to facilitate the binding of regulatory proteins. If this is so, then they may reflect the existence of a fundamental mechanism that allows DNA packaging and gene expression to coexist, i.e. an ability to protect defined DNA regions from nucleosome packaging. Such protection could come about if the assembly of nucleosomes on DNA was not random, but instead regulated so that nucleosomes are assembled on some DNA sequences but not others. This same nucleosome positioning mechanism could be used to block the binding of regulatory proteins by placing a nucleosome on or close to their binding sequences on DNA. In the following section, we examine the evidence that such positioning exists and at the mechanisms by which it is brought about.

Nucleosome positioning *in vitro* and *in vivo*

In an earlier section, we saw how gentle digestion of cell nuclei with micrococcal nuclease cut the DNA into pieces that were multiples of about 200 bp, the exact figure depending on the cell type. This is because the nuclease cuts preferentially in the linker DNA. The spacing of nucleosomes is influenced by the presence or absence of linker histones and possibly their type (H1, H1^0, H5, etc.) and also, in some circumstances, by the presence of histone variants. However, there is increasing evidence that, in some regions of the genome, nucleosome spacing does not tell the whole story. In these regions, nucleosomes have been shown to be positioned precisely with respect to specific DNA sequence elements. (Information on nucleosome positioning has been obtained largely through the use of the indirect end labelling technique described in Box 3.) Note that precise positioning of a few nucleosomes, perhaps even just one, when combined with regular nucleosome spacing, could lead to a situation in which each nucleosome over an extensive chromosome domain has a defined position.

DNA sequence can help determine the position of nucleosomes

To wrap almost two complete turns of DNA around the histone octamer to form a nucleosome core particle requires significant bending of the DNA double helix. This bending is accommodated by compression (narrowing) of the minor groove on the inside of the supercoiled DNA (i.e. the side adjoining the histones) and a matching expansion on the outside. (This was discussed in Chapter 3 and is illustrated in Fig. 3.5.) The difference can be almost two-fold. The ease with which any given piece of DNA can be bent in this way is strongly dependent on its sequence. One of the most clear-cut examples of this is that AT base pairs preferentially occupy inside positions and GC base pairs outside ones, reflecting their relative abilities to accommodate compression of the minor groove. However, the rules that link DNA bendability to sequence are complex. The first attempts to define such rules were based on the finding that the electrophoretic mobility of DNA, under appropriate conditions, is dependent on its resistance to bending. If two pieces of DNA of the same size but different sequence are electrophoresed under the same conditions, the one that is more flexible will migrate faster, presumably because its flexibility enables it to wriggle more efficiently through the supporting medium, usually polyacrylamide. Later work examined the ability of random, synthetic DNA fragments to assemble into nucleosomes *in vitro* using the salt-dialysis procedure. Those DNA molecules that assembled most efficiently were identified by successive rounds of chromatin assembly and isolation, followed by DNA cloning and sequencing. Both approaches showed that certain DNA sequences strongly favour DNA bending and nucleosome formation, whereas others can effectively prevent it. But they also showed that there are no simple rules or sequence motifs governing either process.

These experiments led to an important conclusion, namely that when nucleosomes are assembled on a piece of DNA they will not necessarily be randomly positioned, but will assemble preferentially at sites within that piece of DNA that favour DNA bending. Such sites are described as nucleosome positioning sequences. In some cases the positioning signal is so strong that nucleosomes are assembled with great precision, sometimes down to the single base-pair level. A sequence designed to allow the preferred 'AT inside, GC outside' conformation of nucleosomal DNA provides just such a strong positioning signal. The sequence is a tandem repeat of $W_3N_2S_3N_2$, where W (weak) indicates A or T, S (strong) indicates G or C and N is any nucleotide. With a DNA helical repeat of 10, this sequence automatically places AT pairs on the inside and GC pairs on the outside. Examples of strong, *naturally occurring* positioning sequences are provided by the PolIII-transcribed 5S rRNA genes from several species. These small (120 bp), highly conserved and highly repeated genes have been extensively used for *in vitro* experiments to test the binding of transcription factors to nucleosomes (see Chapter 7).

Synthetic DNAs have proven useful for *in vitro* analysis of the energetics of nucleosome assembly. Competitive reconstitution experiments have been used to test the relative ease with which different DNA sequences can be assembled into nucleosomes. DNA fragments containing the synthetic DNA described above, present in up to five copies, were compared with naturally occurring 5S rRNA genes and mixed, mononucleosomal DNA. The experiments showed that the ability of the DNA to assemble a nucleosome increased in direct proportion to the number of copies of the synthetic positioning sequence. DNA with five copies was incorporated into nucleosomes *in vitro* about 100 times more readily than mixed DNA isolated from native mononucleosomes. The best 5S rRNA positioning sequence was comparable to a fragment with only two copies of the synthetic DNA.

Nucleosomes retain some mobility when positioned by DNA sequence

The nucleosome is held together by a multitude of ionic interactions between the DNA and the core histones. The interactions are individually weak, but collectively they produce a stable particle whose complete dissociation requires high salt concentrations or ionic detergents. However, for the purposes of the present discussion, it is important to know whether this stability is accompanied by equally stable nucleosome positioning.

Very soon after the structure of the nucleosome had been defined, it was shown that it was possible for nucleosomes to move along DNA without dissociating from it, a property described as 'sliding'. However, this movement occurred only at higher ionic strengths and its relevance to events *in vivo* is uncertain. More recently, it has been shown that nucleosomes assembled *in vitro* with a strong positioning sequence (the 5S rRNA gene) can undergo subtle positional changes at low ionic strengths and in the absence of reagents designed to dissociate or weaken DNA–histone interactions. These experiments use an electrophoretic approach similar to the methodology used to detect DNA bending, but in which the nucleosomes themselves are run out on the gel, rather than DNA fragments. The methodology is outlined in Fig. 6.9. There is a close relationship between electrophoretic mobility and the position of the histone octamer on the DNA. Electrophoresis is normally conducted at 4°C to minimize nucleosome rearrangements. If nucleosome-containing regions of the gel are excised, incubated at 37°C for a few minutes and re-electrophoresed, then single bands often split into several bands of differing mobility (Fig. 6.9). It can be concluded from this that incubation at normal body temperature is sufficient to induce movement of the histone octamer along the DNA. This was found to be the case with the 5S rRNA positioning sequence and bulk DNA, although the

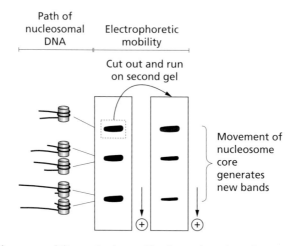

Fig. 6.9 Nucleosome mobility can be detected by electrophoresis. In the experiment shown, the histone octamer can adopt any one of three favoured positions on the DNA (determined by positioning sequences). The three different conformations that result can be separated by electrophoresis, largely as a result of the differing efficiencies with which they thread their way through the gel matrix. Such gels are normally run at low temperature (4°C) and nucleosomes retain their positions. However, if a band is cut out, exposed to higher temperature (e.g. 37°C) and re-run, changes in position can be detected. This technique is useful for measuring the activity of chromatin remodelling enzymes (Chapter 8).

extent of the movement varied. Thus, even a strong, naturally occurring positioning sequence does not confine the nucleosome to just one position, but, rather, defines a small set of positions, of which one is preferred. Nucleosomes can spontaneously shift between these positions.

Perhaps the most significant observation made during these experiments was that the different positions are separated by about 10 bp of DNA, or one complete turn of the DNA helix. In other words, the *rotational* positioning of the DNA is maintained, i.e. base pairs that were originally on the inside remain there, even though the *translational* positioning of the nucleosome has moved by 10 bp. This is illustrated in Fig. 6.10. Exactly how the histone octamer moves (rotates?) within the DNA supercoil through a 10 bp increment remains a mystery, but the fact that such movements can occur is relevant when considering the possible role of nucleosome positioning *in vivo*. For example, movement of a positioned nucleosome through 10, 20 or 30 bp could be enough to reveal a previously protected DNA sequence, and thereby allow transcription factor binding and inappropriate gene expression. These results raise the suspicion that *DNA sequence alone* may not be a strong enough positioning signal to be useful *in vivo*. This is given some support by results showing that DNA sequences that position nucleosomes *in vitro* do not do so *in vivo* when introduced into yeast or *Drosophila* genomes. Perhaps DNA sequence is only one of several factors that are necessary.

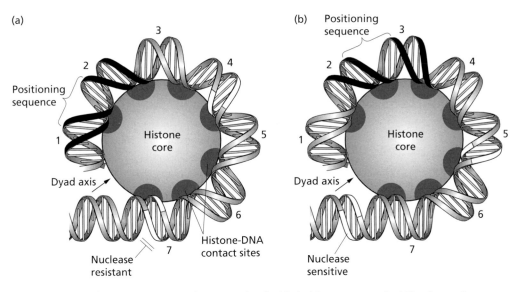

Fig. 6.10 Movement of DNA associated with the histone octamer by 10 bp changes its *translational* positioning but not its *rotational* positioning.

Nucleosomes can be precisely positioned *in vivo*

The *in vitro* mobility of nucleosomes raises a question as to whether positioning is likely to have any relevance to transcriptional control *in vivo*. A search for evidence of nucleosome positioning *in vivo*, usually employing the technique of indirect end labelling (see Box 3, p. 113) has so far revealed over 20 genes in yeast, and a similar number in higher eukaryotes, that show strong evidence of positioned nucleosomes. Some of these have been studied in great detail and have confirmed the important role played by nucleosomes positioned on, or adjacent to, crucial DNA sequence elements. The first example is from the yeast *S. cerevisiae*. (Further examples, described in the next chapter, will introduce some interesting complications.)

The *PHO5* gene

The *PHO5* gene in *S. cerevisiae* encodes an acid phosphatase and is induced when levels of inorganic phosphate (Pi) in the growth medium are very low. Under these conditions there is a 50-fold increase in levels of acid phosphatase secreted by the cells, most of which is pho5. Optimum induction of *PHO5* requires two proteins that bind to sites in the promoter. These are Pho4, a basic helix–loop–helix transactivator that binds to two nonidentical Upstream Activating Sequences (*UASp1* and *UASp2*) and Pho2, a homeobox-containing, DNA-binding protein (Fig. 1.6a) that binds to several sites in the promoter, one of which overlaps with *UASp1* and two of which flank *UASp2*. Pho4 is regulated by protein kinases that lead to its phosphorylation (and inactivation) under high-phosphate conditions. The most remarkable feature of the *PHO5* promoter region is that it is packaged into an array of four, precisely positioned nucleosomes. These nucleosomes obscure the TATA box and *UASp2* but not *UASp1*, which is located in a short, nucleosome-free region hypersensitive to nuclease digestion. This is shown diagrammatically in Fig. 6.11.

Induction of the *PHO5* gene leads to rapid and dramatic loss of nucleosome positioning detected by nuclease digestion and end-labelling. It is important to note that the results do not necessarily mean that histones are displaced totally from the induced *PHO5* promoter. This could have happened, but all we can be sure of is that the nucleosomes have been altered in such a way that the DNA is more accessible to nucleases. However, we can be confident that the nucleosomes play an important part in the regulation of *PHO5*. As described earlier, in yeast strains in which nucleosome density is reduced by limiting histone synthesis (Fig. 6.6), the *PHO5* gene is transcribed even when Pi concentration is high, although not to the maximal level achievable in wild type cells.

The disruption of chromatin structure that accompanies *PHO5* activation is not a simple consequence of transcription. Nucleosome reorganization still occurs at low Pi concentrations in cells carrying a mutated *PHO5* TATA box and in

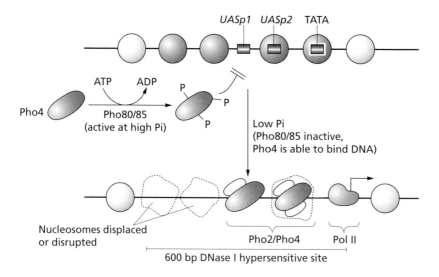

Fig. 6.11 Positioned nucleosomes on the *PHO5* promoter are displaced in the presence of low concentrations of inorganic phosphate (Pi). The transcription factor Pho4 is phosphorylated by kinases that are active under high Pi conditions. Phosphorylated Pho4 cannot bind DNA. The kinases are inactive under low Pi conditions, with the result that Pho4 becomes dephosphorylated and able to bind DNA (along with Pho2) and initiate transcription. The four densely shaded nucleosomes are all changed during the activation process so as to render the DNA associated with them nuclease sensitive, but whether they are completely displaced or structurally modified remains unclear.

which the gene cannot be transcribed. Nor is DNA replication required for *PHO5* activation. What *is* essential is binding of Pho4 to the nucleosome-free site *UASp1*. This allows binding of a second Pho4 molecule to the nucleosomal *UASp2* site (see Fig. 6.11). Under normal conditions, there is no binding of Pho4 to *UASp2* when *UASp1* is mutated. However, overproduction of Pho4 can 'force' binding to *UASp2*, showing that the nucleosome is not an *absolute* barrier. Significantly, under these conditions chromatin disruption is the same as in the wild type promoter. This suggests that disruption of nucleosomes 1–4 is an all-or-nothing event that, once triggered, proceeds to completion. The molecular mechanism that brings about this chromatin reorganization remains unclear, although the chromatin remodelling enzymes discussed in Chapter 8 are likely to be involved.

It is interesting to ask why the *PHO5* promoter makes life so complicated. Would it not be much simpler just to have activated Pho4 bind to a UAS and activate the gene? What is gained by having a second UAS, and the TATA box, buried in nucleosomes? The answer may lie in the observation that *UASp2*, when nucleosome-free, can bind an alternative transcriptional activator, Cpf1. This activator is related to Pho4, but is differently regulated and has a different function. Hiding *UASp2* in a nucleosome will prevent binding of Cpf1, and possibly other related proteins, and the consequent inappropriate activation of the *PHO5* gene. There is also no doubt that hiding the TATA box in a nucleosome is a very effective way of preventing adventitious binding of TATA binding protein (TBP, see Chapter 7), something that would also bring about inappropriate activation.

Chromatin domains

There is no such thing as a standard chromosome domain. There are properties and characteristics that are often present in domains, but no domain has all of them. Perhaps the best definition of a domain is a functional one, namely, it is a region of the genome whose constitutent genes are regulated independently of the genes or chromosome regions that surround them. Domains are generally of the order of 50–200 kb in size, and are often characterized by a relatively high or relatively low level of nuclease sensitivity. Regions of high sensitivity are usually associated with transcriptionally active or potentially active genes, and in some cases will encompass a group of functionally related genes. A model (minimal) domain is shown in Fig. 6.12. It contains two genes, with promoters, proximal transcription factor binding sites and an enhancer, several kb further upstream, that regulates transcription from both genes. The domain is transcriptionally active, despite the fact that the chromatin surrounding it is condensed (nuclease resistant). In order to prevent the surrounding chromatin encroaching on the domain and silencing it, or to prevent spreading of the 'active' structure beyond the domain, the domain is protected at its 5' and 3'

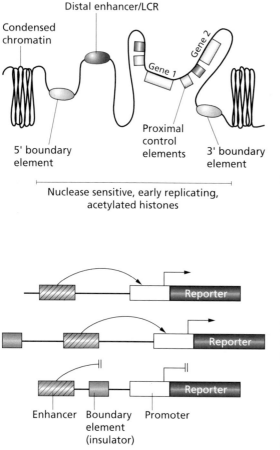

Fig. 6.12 Composition of a hypothetical chromatin domain.

Fig. 6.13 Boundary elements can sometimes prevent interaction between enhancers and promoters. The diagram shows three DNA constructs used to test boundary element function by transfection into cultured cells.

ends by 'boundary elements'. These DNA sequences are also known as insulators and are defined operationally by their ability to block the interaction between enhancers and promoters, usually in transient transfection experiments (Fig. 6.13). More than 10 insulator sequences have now been identified in a variety of species. There is no simple consensus sequence that is shared by these insulators, but it seems likely that they all operate through binding of specific, perhaps different, proteins. Given that we really have very little idea about how promoters and distant, upstream enhancers interact in order to increase transcription, thoughts about the way in which insulators inhibit this interaction are bound to be speculative. What is remarkable is that the sequences seem to be able to operate across vast evolutionary distances. Thus, a 1200 bp insulator from the *Drosophila* bithorax complex, Fab-7, can operate very effectively in mammalian cells.

Perhaps the best-studied example of a chromatin domain, at least in vertebrates, is the β-globin locus. This locus contains a family of genes that makes proteins responsible for oxygen transport in vertebrate red cells. (The α-globin subunits are made by a separate, independently regulated gene.) The locus has been highly conserved through vertebrate evolution. The human version is shown in Fig. 6.8. The different genes are expressed at different times through embryonic development, as indicated, in order to provide haemoglobin molecules best suited to the demands of the embryo, the foetus and the adult. The domain is characterized by a cluster of DNase hypersensitive sites (DHS) upstream of the ε-globin gene. Collectively, these sites make up a rather special type of enhancer element called the locus control region (LCR). The properties of the human LCR are understood mainly as a result of experiments in which the human β-globin locus has been stably inserted into the genome of transgenic mice. Remarkably, if the locus is intact, the various human genes are expressed at the correct developmental stage in their mouse hosts. Mutations in different regions can then be used to study how expression is regulated. If some or all of the DHS comprising the LCR are missing, then the locus suffers the fate that befalls most transgenes, i.e. its transcription is suppressed in an unpredictable way that depends on exactly where in the host genome it has been inserted. This phenomenon (position effect variegation) is dealt with in a later chapter. For now, the important point is that the LCR prevents this happening. It has the ability both to open chromatin structure across the locus and to protect it from the effects of surrounding chromatin. An element with the properties of an insulator has been located to 5′ HS5.

The chromatin structure of the native β-globin locus has been studied closely in the chicken, a vertebrate whose red blood cells retain their nuclei (unlike those of mammals) and have been used extensively as a source of relatively homogeneous chromatin. The DNaseI sensitive domain was defined and shown to coincide exactly with a region of reduced histone H1 levels and increased histone acetylation, measured by chromatin immunoprecipitation. It is this experiment that has given rise to the idea that DNaseI sensitivity and elevated levels of histone acetylation are coincident. Whether this coincidence is a general phenomenon or confined only to some loci remains to be seen. A 1.2 kb DNA fragment from the chicken LCR, specifically from HS4, has all the properties of an insulator, preventing enhancer action in human cell lines and protecting against position effects in *Drosophila*. Its highly conserved insulating properties seem to derive from a GC-rich fragment only 250 bp long. The 3′ insulator remains to be identified.

Further Reading

Nucleosomes, chromatin and transcription

Ericsson, C., Grossbach, U., Björkroth, B. & Daneholt, B. (1990) Presence of histone H1 on an active Balbiani ring gene. *Cell*, **60**: 73–83.

Ericsson, C., Mehlin, H., Bjorkroth, B., Lamb, M. M. & Daneholt, B. (1989) The ultrastructure of upstream and downstream regions of an active Balbiani ring gene. *Cell*, **56:** 631–639.

Han, M. & Grunstein, M. (1988) Nucleosome loss activates yeast downstream promoters *in vivo*. *Cell*, **55:** 1137–1145.

Solomon, M. J., Larsen, P. L. & Varshavsky, A. (1988) Mapping protein–DNA interactions *in vivo* with formaldehyde: evidence that histone H4 is retained on a highly transcribed gene. *Cell*, **53:** 937–947.

Wu, C., Wong, Y.-C. & Elgin, S. C. R. (1979) The chromatin structure of specific genes: II. Disruption of chromatin structure during gene activity. *Cell*, **16:** 807–814.

Nucleosome positioning and mobility

Drew, H. R. & Calladine, C. R. (1987) Sequence-specific positioning of core histones on an 860 base-pair DNA. Experiment and theory. *J. Mol. Biol.*, **195:** 143–173.

Meersseman, G., Pennings, S. & Bradbury, E. M. (1992) Mobile nucleosomes—a general behaviour. *EMBO J.*, **11:** 2951–2959.

Simpson, R. T., Roth, S. Y., Morse, R. H. *et al.* (1993) Nucleosome positioning and transcription. *Cold Spring Harbor Symp. Quant. Biol.*, **58:** 237–245.

Svaren, J. & Hörz, W. (1995) Interplay between nucleosomes and transcription factors at the yeast *PHO5* promoter. *Semin Cell Biol.*, **6:** 177–183.

Nuclease sensitivity and hypersensitive sites

Gross, D. S. & Garrard, W. T. (1988) Nuclease hypersensitive sites in chromatin. *Ann. Rev. Biochem.*, **57:** 159–197.

Stalder, J., Larsen, A., Engel, J. D. *et al.* (1980) Tissue-specific DNA cleavages in the globin chromatin domain introduced by DNase I. *Cell*, **20:** 451–460.

Chromatin domains, control regions and boundaries

Bell, A. C. & Felsenfeld, G. (1999) Stopped at the border: boundaries and insulators. *Curr. Opin. Genet. Dev.*, **9:** 191–198.

Bender, M. A., Bulger, M., Close, J. & Groudine, M. (2000) β-*globin* gene switching and DNase I sensitivity of the endogenous β-*globin* locus in mice do not require the Locus Control Region. *Mol. Cell*, **5:** 387–393.

Hebbes, T. R., Clayton, A. L., Thorne, A. W. & Crane-Robinson, C. (1994) Core histone hyperacetylation comaps with generalized DNase I sensitivity in the chicken β-globin chromosomal domain. *EMBO J.*, **13:** 1823–1830.

Kellum, R. & Elgin, S. C. R. (1998) Chromatin boundaries: punctuating the genome. *Curr. Biol.*, **8:** R521–524.

Prioleau, M.-N., Nony, P., Simpson, M. & Felsenfeld, G. (1999) An insulator element and condensed chromatin region separate the chicken β-globin locus from an independently regulated erythroid-specific folate receptor gene. *EMBO J.*, **14:** 4035–4048.

Schübeler, D., Francastel, C., Cimbora, D. M. *et al.* (2000) Nuclear localization and histone acetylation: a pathway for chromatin opening and transcriptional activation of the human β-globin locus. *Genes Dev.*, **14:** 940–950.

Strouboulis, J., Dillon, N. & Grosveld, F. (1992) Developmental regulation of a complete 70-kb human β-globin locus in transgenic mice. *Genes Dev.*, **6:** 1857–1864.

Nuclear domains and action at a distance

Bulger, M. & Groudine, M. (1999) Looping vs. linking: towards a model for long-distance gene activation. *Genes Dev.*, **13**: 2465–2477.

Ferreira, J., Paolella, G., Ramos, C. & Lamond, A. I. (1997) Spatial organization of large-scale chromatin domains in the nucleus: a magnified view of single chromosome territories. *J. Cell Biol.*, **139**: 1597–1610.

Chapter 7: How the Transcription Machinery Deals with Chromatin

Introduction

Nucleosomes can be assembled *in vitro* using defined pieces of DNA and purified histones. By incorporating DNA sequences that act as promoters or that are recognized by particular DNA-binding proteins, these *in vitro* constructs become invaluable for studying the role of the nucleosome in transcription or transcription factor binding. If a strong nucleosome positioning sequence is also built into the DNA used for assembly, the histone octamer can be placed at a specific site on the DNA. So, simple *in vitro* systems have a lot of advantages. Conversely, it is worth bearing in mind that the nucleosome core particle, *in vivo*, is always part of a higher-order chromatin structure. So, while the behaviour of core particles *in vitro* points to what is physically possible, it is not necessarily a reliable indicator of what really happens in the cell nucleus. *In vitro* systems of increasing complexity that are designed to model more closely the situation *in vivo* will be examined later in the chapter.

In vitro studies of transcription factor binding

Experiments to test the ability of DNA-binding proteins to bind nucleosomal DNA start with the assembly of nucleosomes *in vitro*. This is most simply done by salt gradient dialysis of histones and pieces of DNA containing one or more binding sites for the protein being studied and a nucleosome positioning sequence. A pure preparation of the protein under test is mixed with the self-assembled, purified nucleosomes, together with any potential auxiliary factors. Protein binding is assessed by electrophoresis in nondenaturing gels. A nucleosome to which a protein has bound will migrate more slowly through the gel than one to which it has not. That the bound protein is really the one we are interested in can be confirmed by transferring proteins from the gel to a filter and labelling with antibodies to the protein of interest. Alternatively, we can add antibodies before electrophoresis and look for additional slowing ('supershifting'). This approach is summarized in Fig. 7.1.

Despite describing this as a 'simple' system, experiments of this type are technically difficult. The importance of experimental variables is apparent from the fact that some proteins have shown rather different binding behaviour when tested in different laboratories. It is worth spending a little time considering the potential pitfalls of this type of experiment, if only to define what questions we should ask in assessing the results and how cautious we should be in interpreting them.

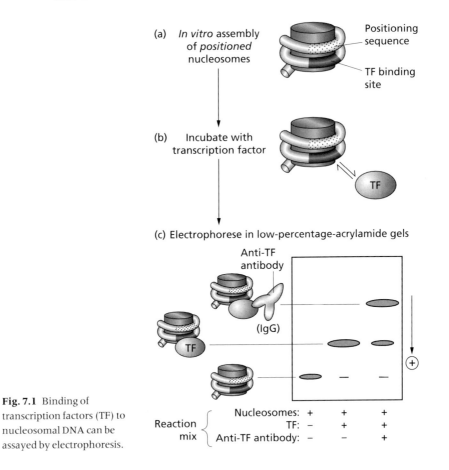

Fig. 7.1 Binding of transcription factors (TF) to nucleosomal DNA can be assayed by electrophoresis.

Firstly, how homogeneous (or otherwise) are the assembled nucleosomes? Histones used for assembly have usually been purified from cells or tissues and will be a heterogeneous mixture of isoforms differing in the level of post-translational modification (e.g. acetylation) and of histone variants. This problem will disappear as the use of *recombinant* histones for chromatin assembly becomes increasingly common. Secondly, how strong is the positioning sequence that determines translational and rotational positioning? Methods outlined in the previous section should be used to define, accurately, the proportion of nucleosomes in which the DNA is 'correctly' positioned. As noted earlier, there is ample evidence that nucleosome positioning is not a fixed characteristic, but instead represents an equilibrium between a limited number of alternative positions around a particular point. If this is the case, then the protein under test may be able to bind to the nucleosome when, by chance, it adopts a position favourable for binding. If the protein then remains stably bound, a high level of bound protein will accumulate progressively, even though only a small minority of nucleosomes is actually capable of binding at any one time. Finally, how dependent are the results on ionic conditions during the protein–nucleosome binding step? It is an unfortunate fact that chromatin *in vitro* is not stable under the ionic conditions that approximate what we believe to be the situation *in vivo*. Small changes in, for example, the concentration of divalent cations can alter chromatin structure in functionally significant ways.

Despite all these technical complexities, *in vitro* binding experiments have provided, and continue to provide, important results that give clues about what may be happening *in vivo*. These results may be summarized as follows:

1 The nucleosome generally inhibits the binding of proteins to DNA, although there are important exceptions. For some proteins, such as the general transcription factor NF1, the inhibition is such as to prevent any functionally significant binding, while for others, such as the yeast transcriptional activator GAL4, significant binding is still possible.

2 Translational and rotational positioning can influence binding. By adjusting the position of the nucleosome positioning sequence relative to the factor binding site, both rotational and translational positioning can be altered. GAL4 is sensitive to translational position in that it binds more easily to sites near the points at which DNA enters and leaves the nucleosome than to sites near the dyad axis (Fig. 7.2). Presumably, it is easier for the protein to peel the DNA away from the histone octamer at the end rather than in the middle. Steroid receptors are particularly sensitive to rotational positioning, binding only when the response element is located on the side of the DNA helix away from the histone core (Fig. 7.3). The nuclease DNAseI has the same property. In contrast, NF1 cannot be induced to bind by changing either translational or rotational positioning, at least until its binding site leaves the nucleosome altogether. It is likely that the inability of NF1 to bind nucleosomal DNA is a consequence of its mode of binding. NF1 binds with high affinity and makes contacts with numerous sites along the DNA (over 20 have been identified). In doing so, it almost surrounds the DNA helix. This type of binding is impossible when one face of the target DNA is bound to a protein surface. The situation is similar to that which accounts for the inability of micrococcal nuclease to cleave nucleosomal DNA (see Fig. 3.4). In contrast, proteins such as steroid receptors (discussed further below) bind with lower affinity (i.e. make fewer contacts) to only one face of the DNA and can therefore attach themselves to nucleosomal DNA, provided that the binding site is located on the outer surface.

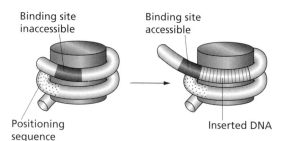

Binding site
inaccessible

Binding site
accessible

Positioning
sequence

Inserted DNA

Fig. 7.2 Changes in the *translational* positioning of nucleosomal DNA can alter transcription factor binding.

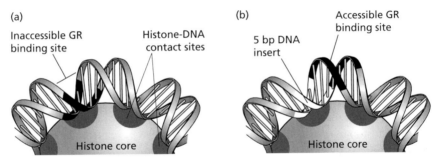

Fig. 7.3 Changes in the the *rotational* positioning of nucleosomal DNA can alter accessibility of the glucocorticoid receptor binding site. (a) A positioning sequence locates the DNA, relative to the histone octamer, in such a way that the binding site for the glucocorticoid receptor (GR) is located on the inner face of the DNA and therefore inaccessible to the protein. (b) If a piece of DNA five bases long is inserted in front of the GR binding site, its rotational positioning is changed so that it is now located on the outer face of the DNA. The rotational positioning of the positioning sequence remains the same.

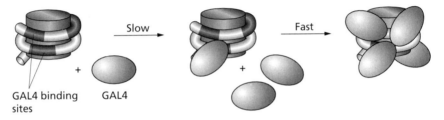

Fig. 7.4 Binding of GAL4 to multiple sites on nucleosomal DNA is cooperative.

3 Bound proteins can modify the nucleosome so as to facilitate subsequent protein binding. In some cases, this can lead to cooperative binding to nucleosomes containing more than one binding site for the same protein. For example, binding of GAL4 to nucleosomes constructed so as to contain four GAL4 binding sites is strongly cooperative (i.e. the binding of one GAL4 molecule facilitates binding of another, Fig. 7.4). Such cooperativity is not seen when the same piece of DNA is tested without being assembled into a nucleosome. Such cooperative effects provide a possible reason for situations in which two or more adjacent binding sites for regulatory proteins are found *in vivo*. Examples include the Steroid Response Elements (SRE) of steroid-dependent promoters such as that of the Mouse Mammary Tumour Virus described below. Significantly, experiments in which nucleosomes were constructed with DNA containing binding sequences for GAL4 and unrelated transcription factors (such as the mammalian factor NF-κB) showed that cooperativity occurred with all combinations of factors tested. This result makes it difficult to invoke specific protein–protein interactions as the source of the cooperative binding effects observed *in vitro*. Instead, binding of the first factor must change the nucleosome in such a way that binding of subsequent factors is facilitated. In this respect, the cooperativity involved here is different from that occurring in what is perhaps the best known example of the cooperative binding of proteins to DNA, namely binding of the λ repressor protein to three adjacent operator sites in bacteriophage λ. In this case, the cooperativity is brought about primarily through protein–protein interactions.

4 Histones are not generally displaced by the binding of proteins to nucleosomal DNA, although histone–DNA interactions may be weakened. For example, although histones are not lost immediately from the nucleosome by the attachment of multiple GAL4 molecules, addition of competing histone-binding polycations to the reaction mix leads to progressive nucleosome displacement, showing that the equilibrium between bound and unbound histones is shifted perceptibly towards the unbound side.

5 Removal of the histone N-terminal tails or their modification by acetylation can influence factor binding. The N-terminal tail domains of the core histones are exposed on the nucleosome surface and (unlike the much shorter C-terminal domains) are sufficiently long that their presence might influence the binding of proteins to nucleosomal DNA. Results have been presented to suggest that binding of the transcription factor TFIIIA to the *Xenopus* 5S RNA gene and binding of GAL4 in the system outlined above can be influenced by both the presence or absence of histone tails and by their state of acetylation. In both cases, proteins bound more efficiently to nucleosomes in which the N-terminal tails were absent or highly acetylated (Fig. 7.5). On steric grounds, this is the result that one might have predicted, but unfortunately the findings with TFIIIA have proved difficult to reproduce, possibly as a result of technical factors of the type discussed earlier. The role of histone tails in transcription factor binding remains controversial. But, to put this in perspective, it is important to remember that, *in vivo*, the tail domains may well be constrained by association with other proteins or with specific regions of DNA. At this level of detail, *in vitro* experiments may not tell us a great deal about what actually happens *in vivo*.

Some energetic considerations

The binding energy of an average transcription factor to its cognate DNA sequence is around 12–15 kcal mol^{-1}. In comparison, binding of DNA to the

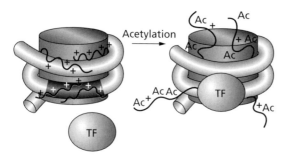

Fig. 7.5 Acetylation (or removal by proteolysis) of histone N-terminal tail domains can expose transcription factor (TF) binding sites and facilitate TF binding.

histone core is governed by weak, electrostatic forces, amounting to about $0.1–0.15\,kcal\,mol^{-1}$ bp at 0.1 M salt. (The strength of DNA:histone octamer binding is very sensitive to ionic strength.) The figures, approximate though they are, show that the displacement of up to 10 bp of DNA from the histone core to allow a protein to bind is likely to be energetically favourable. Cooperative binding of the type shown by GAL4 may be a consequence, at least in part, of the progressive dissociation of DNA from the nucleosome core so as to make binding sites more accessible. This may be initiated by binding first at positions where the DNA is entering and leaving the nucleosome and where histone–DNA interactions will be limited.

A crowded nucleosome: Mouse Mammary Tumour Virus nucleosome B

The promoter of the Mouse Mammary Tumour Virus (MMTV) has proved to be an extremely popular and informative model system with which to study the binding of various transcription factors to chromatin. Multiple copies of the MMTV genome are integrated into the DNA of most inbred mouse strains. As is common with viruses of this type, MMTV uses control elements originally hijacked from the cell itself to regulate transcription of its own genes. The upstream long-terminal repeat of MMTV contains binding sites for various eukaryotic transcription factors together with steroid response elements (SREs) — sites that allow binding of ligand-bound steroid hormone receptors. Thus, viral genes will be expressed only in the presence of hormone and only in cells that contain steroid hormone receptor proteins. MMTV DNA can be isolated and inserted, in single copy, into selected cells to provide an extremely valuable model system for studying the operation of mammalian hormone response elements.

The factor binding sites in the MMTV promoter are shown in Fig. 7.6. There are four binding sites for the steroid hormone receptor (SR) protein (TGTTCT motifs), together with sites for the OCT1 and NF1 transcription factors. All of these factors must bind in order to achieve maximal transcription. The exact stoichiometry of binding of steroid receptors to the SREs is unclear, although mutation of any one site causes a major reduction in transcription initiation and there is evidence for strong cooperativity between the different sites. The symmetrical motif (TGTTCTnnnAGAAGA), together with adjacent bases, will bind the liganded (i.e. hormone-bound) glucocorticoid receptor as a homodimer and much experimental work has been done with this receptor. However, the same sites will also bind some other members of the steroid hormone receptor superfamily, hence their more general designation as SREs.

The nuclease digestion/end-labelling procedure outlined in the previous chapter has been used to show that, *in vivo*, the MMTV promoter is characterized by six positioned nucleosomes (Fig. 7.6). One of these, nucleosome B, incorporates all four SREs, the NF1 site and possibly one of the two OCT1 sites.

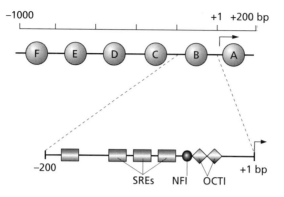

Fig. 7.6 Nucleosomes are positioned across the promoter region of the Mouse Mammary Tumour Virus (MMTV). Nucleosome A obscures the transcription start site. Nucleosome B incorporates four steroid response elements (SREs), sequences that are recognized and bound by the (liganded) glucocorticoid (or progesterone) receptor proteins, along with NF1 and OCT1 sites.

The TATA box falls within, or adjacent to, nucleosome A. This positioning is seen in copies of MMTV grown as multicopy episomal (replicating) plasmids in host cells, or incorporated, as a single copy, into their chromosomes. The same positioning has been found in chromatin assembled *in vitro* using fragments of the MMTV LTR, suggesting that the positioning signal lies in the DNA sequence. Exactly what this signal is, or how it is interpreted during chromatin assembly, remains unclear.

Single-nucleosome binding studies of the sort described earlier have shown that glucocorticoid receptor (GR) binding is significantly reduced when its cognate sequence is incorporated into a nucleosome, and NF1 binding is essentially abolished. For these reasons alone, nucleosome B must exert a significant influence on the expression of genes driven by the MMTV promoter. *In vitro* binding experiments have also shown that the rotational positioning of nucleosomal DNA has a crucial effect. Thus, when the sequence recognized by GR is on the inside of the DNA helix (i.e. adjacent to the nucleosome surface), binding is much less than when it is on the outside (Fig. 7.3). This observation introduces a further subtle mechanism by which chromatin can influence transcription, namely by slight shifts in nucleosome positioning causing changes in the rotational positioning of a factor binding site. Nucleosome B is frequently positioned *in vivo* such that two of the SREs (II and III) are on the inside and not available for binding, and the other two are on the outside.

A second important result from MMTV studies is that GR binding facilitates subsequent binding of two additional factors, namely NF1 (which cannot bind on its own) and OCT1 to give a nucleosome packed with bound factors. *In vitro* binding studies have shown that these factors cannot all bind together on naked DNA; only when DNA is folded on the nucleosome can all the sites be occupied simultaneously. This is shown in Fig. 7.7. It provides another example of a situation where chromatin provides a *permissive* environment for factor binding. The exact disposition of the different factors, or their stoichiometry, is not known, but it is clear that the histones are not displaced. This is an important result, because this crowded nucleosome has a site close to its dyad axis that is extremely sensitive to digestion by nucleases (i.e. a DNaseI hypersensitive site). The presence of this site gave the impression at first that the region occupied by steroid receptors and other factors was nucleosome free. It clearly is not.

It remains an open question whether the chromatin remodelling that accompanies binding of ligand-bound GR to nucleosome B is brought about entirely by the synergistic binding of the various transcription factors. It may also require some help from enzymes that can remodel chromatin, such as SWI/SNF, or that can modify histones, such as histone acetyltransferases. (These enzymes are discussed in Chapter 8.)

The MMTV story has been complicated somewhat over recent years by the realization that the nucleosome positions outlined in Fig. 7.6 are more variable than was originally supposed. A technique involving nuclease digestion of formaldehyde-fixed chromatin and characterization of mono-nucleosome-sized fragments by primer-extension showed that nucleosome B has three common (i.e. preferred) positions, and several less common ones, all within about 50 bp of one another. The major positions are marked in Fig. 7.8. It is interesting that the favoured locations fall at intervals of about 10 bp, i.e. one turn of the DNA helix, so *rotational* positioning is maintained from one location to the next. As the multiple locations were detected by analysis of chromatin cross-linked *in vivo* by treatment with formaldehyde, we can conclude that the variability is not a result of nucleosomes sliding around during chromatin isolation. It is not attributable to the sort of *in vitro* nucleosome mobility discussed earlier. However, nucleosomes may be mobile *in vivo* within individual cells, oscillating between

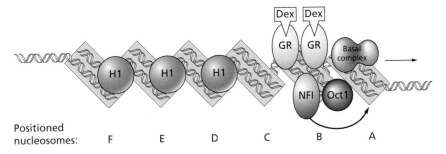

Positioned nucleosomes: F   E   D   C   B   A

Fig. 7.7 Activation of the MMTV promoter leads to assembly of multiple transcription factors at the DNA sites occupied by nucleosomes A and B. The accumulation of transcription factors leads to major alterations in nucleosome structure and the appearance of DNaseI hypersensitive sites, but does not seem to involve complete displacement of the histone octamer. (Redrawn from Smith and Hager, 1997.)

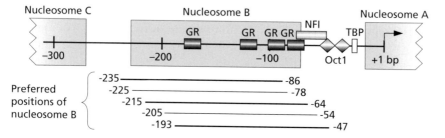

Fig. 7.8 Nucleosome B has a set of preferred positions on MMTV promoter DNA, rather than a single fixed position. The heavy lines represent the major positions and the lighter lines ones that are used less often. Note that the positions are located at about 10 bp intervals, so the *rotational* positioning of nucleosomal DNA will remain much the same.

one preferred position and the next. Alternatively, different locations may be adopted in different cells and then maintained stably, perhaps until the next round of DNA replication and chromatin assembly. In this context, it may be relevant that if cultured cells carrying chromosomal MMTV constructs are treated with glucocorticoids, only 15–20% of cells will initiate transcription from the MMTV promoter. This contrasts with other promoters regulated by positioned nucleosomes. For example, inducing conditions (i.e. low Pi) will result in 95–100% of yeast cells upregulating *PHO5* expression. In the case of MMTV, it may be that the difference between expressing and nonexpressing cells is a reflection of the variation in nucleosome positioning across the MMTV promoter. Perhaps only some positions allow transcription to be initiated *in vivo*.

In concluding the MMTV story, we could reasonably ask the same question that we asked about the *PHO5* promoter in yeast. Why make things so complicated? In a way, the answer is easier to come up with in this case, because induction of gene expression by steroid hormones is intrinsically complex. Steroid hormones exert multiple physiological effects on their target cells through induction of a variety of genes. The increases in gene expression brought about by liganded steroid receptors are closely regulated and often transient. This on/off effect is modelled exactly by the MMTV promoter. Thus, when cells containing chromosomal or episomal MMTV and expressing steroid hormone receptors are exposed to hormone, there is an up-regulation of transcription from the MMTV promoter within about one hour. Within 24 h, transcription returns to baseline levels, even in the continued presence of hormone. However, in cells in which MMTV promoter constructs are transfected *transiently*, and therefore not packaged into nucleosomes, expression continues indefinitely in the presence of hormone and protein receptor. This finding implicates chromatin in the physiologically essential suppression of transcription. It is also worth noting that the MMTV promoter will respond to various steroids in addition to glucocorticoids. These include androgens, mineralocorticoids and others. At least some of the protein receptors binding these hormones bind to the same TGTTCT sequence motifs as the glucocorticoid receptor. Chromatin may well play a role in determining the specificity of response to these different hormone receptors in different cell types.

Strategies for dealing with the repressive effects of chromatin

One can think of three general ways in which the transcription initiation machinery might overcome, or circumvent, chromatin inhibition *in vivo*. The first is by simply making sure that the site in question is maintained in a nucleosome-free state, usually detected as a DNaseI hypersensitive site. We have already discussed ways in which this might be brought about. A second is by using specialized, enzymatic mechanisms that reorganize chromatin so as to allow factor access. Such mechanisms are discussed in Chapter 8. A third possibility is to take advantage of a time in the cell's life cycle during which the DNA is naturally dissociated from histones, i.e. during or shortly after DNA replication. If this is to be a useful *in vivo* mechanism, then the transcription factors that bind DNA during this post-replication window must remain associated with it during the subsequent chromatin assembly process. This can be tested *in vitro*.

Many *in vitro* experiments use chromatin assembled by the salt gradient dialysis procedure. This has the big advantage of allowing chromatin to be assembled from purified components. However, the disadvantage is that the high salt concentrations required inhibit binding of transcription factors to an even greater extent than that of the histones themselves. By the time salt concentrations have fallen sufficiently to allow TF binding, the chromatin will already have assembled. This technical problem can be circumvented by using nuclear extracts with the ability to assemble chromatin under near-physiological conditions. These extracts are often made from *Xenopus* eggs or oocytes, cells that contain all the material, including histones, necessary for the very rapid rounds of DNA replication and chromatin assembly that follow fertilization. An alternative procedure uses extracts from early *Drosophila* embryos, which are also preoccupied with DNA replication and chromatin assembly rather than transcription. Plasmid DNA added to such extracts will be packaged efficiently into chromatin with properties, such as nucleosome spacing, that are indistinguishable from native chromatin. Furthermore, if transcription factors are added at the same time as the plasmid DNA, their ability to compete with histones for DNA-binding sites can be assessed. Experiments to test this have consistently shown that factors added prior to or during chromatin assembly can effectively alleviate inhibition. An outline of the general approach is shown in Fig. 7.9. The experiment shown uses a plasmid carrying the adenovirus major late promoter (a PolII promoter with a typical TATA box) as a test system. Addition of the TATA-binding complex TFIID, or of the TATA-binding component TBP alone, during chromatin assembly, but not afterwards, reduced inhibition of *in vitro*

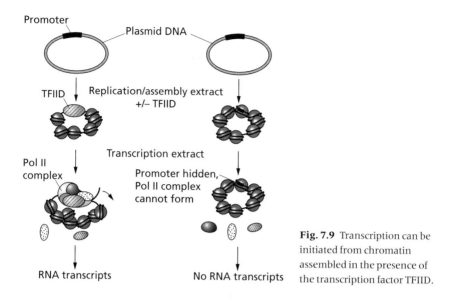

Fig. 7.9 Transcription can be initiated from chromatin assembled in the presence of the transcription factor TFIID.

transcription (measured by incorporation of radiolabelled UTP). The same result is seen with PolIII promoters and those lacking a canonical TATA box, although the factors required differ. For example, efficient transcription of the chromatin-packaged 5S ribosomal RNA genes by PolIII requires the preassembly of a complete transcription complex containing the factors TFIIIA, TFIIIB and TFIIIC. It is interesting to note that TFIIIA alone, while perfectly capable of DNA binding, is not sufficient to prevent inhibitory chromatin assembly. Similarly, transcription from the TATA-less chicken β-globin promoter requires prebinding of *both* the erythroid-specific transcription factors GATA-1 and NF-E4.

Many experiments of this general type have been carried out using a variety of chromatin assembly and transcription systems and transcription extracts. Despite some technical pitfalls, two general rules have emerged. First, the transcription machinery itself does not initiate transcription from promoters packaged as chromatin. Second, prebinding of factors such as TBP and others, sometimes in combination, can put down a marker that survives chromatin assembly, facilitates subsequent assembly of a transcription initiation complex and thereby alleviates the inhibition of transcription that would otherwise occur. This introduces a potentially important *in vivo* mechanism, one that could be used not only for the readying of specific promoters for transcription during DNA replication, but also, potentially, for the maintenance of that transcriptional potential through subsequent rounds of DNA replication and chromatin assembly.

Both the *Xenopus* and *Drosophila* chromatin assembly extracts are naturally deficient in histone H1, so the inhibitory effect observed can be attributable to nucleosome core particles alone. Histone H1 is a major chromatin component in adult cells, and attempts have been made to use the *in vitro* chromatin assembly systems to measure its effect on transcription. In general, such experiments have shown that H1 can increase the chromatin-induced inhibition of transcription, but to a variable extent. Neither result is surprising. As noted in Chapter 3, histone H1 has a central role in the packaging of chromatin into higher-order structures and a variety of studies suggest that its role in transcription, if any, is likely to be inhibitory. The variability of the H1 effects observed is likely to reflect the complexity of the *in vitro* system being used and the many factors that can influence the final result. For example, the density of nucleosomes along the plasmid, the regularity with which they are spaced, the extent to which the histones are modified (e.g. by acetylation) and the levels of other potentially influential nonhistone proteins in the assembly extract will all influence the final levels of transcription.

Nucleosomes occasionally enhance transcription

The emphasis so far has been very much on the ways in which chromatin represses transcription, and the weight of evidence shows that this is by far its most usual effect. However, there are examples in which chromatin, or more specifically a nucleosome, can actually enhance transcription. The first of these came from detailed analyses of nucleosome positioning upstream of the *Drosophila hsp26* heat shock gene by Sarah Elgin and colleagues. A positioned nucleosome was identified between two DNaseI hypersensitive sites separated by about 300 bp. Both of these sites contained DNA sequences that bind heat shock factor (HSF), the protein necessary for the rapid up-regulation of heat shock genes in response to stress (Fig. 7.10). The effect of the positioned nucleosome will be to bring together the two clusters of HSF binding sites, facilitating protein–protein interactions and synergistic binding. Interestingly, positioning of this nucleosome depends not only on DNA sequence, but also on the presence of binding sites for the protein GAGA factor. It seems that GAGA factor plays a role in positioning this nucleosome.

Another example in which a nucleosome exerts a positive effect on transcription comes from *in vitro* studies of the *Xenopus* vitellogenin gene. In this case, DNA looping brought about by a positioned nucleosome brings together a promoter-proximal enhancer element (about 300 bp upstream of the start site), the TATA box and adjacent protein binding sites. The positioned nucleosome gives a 10–20-fold *increase* in transcription compared with naked DNA. In contrast, nonspecific (i.e. nonpositioned) chromatin assembly gives a strong repressive effect.

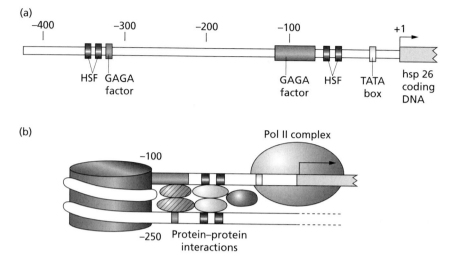

Fig. 7.10 Juxtaposition of protein binding sites on DNA by a positioned nucleosome facilitates protein–protein interactions and initiation of transcription. Transcription of the Heat Shock gene *hsp26* is regulated by several upstream binding sites for the heat shock factor (HSF) and GAGA factor. Binding of these proteins to DNA is facilitated by protein–protein interactions, but on free DNA the distance between the two sets of sites is too great to allow this to happen (a). However, if a nucleosome is inserted between the sites, as shown in (b), the binding sites are juxtaposed on DNA entering and leaving the nucleosome, and a protein complex forms, leading to TBP binding and transcription complex assembly.

The opportunities presented by DNA replication

DNA replication inevitably brings about the disassembly of the chromatin struc-tures within which DNA is packaged for most of the cell cycle. For a relatively brief period, short stretches of DNA are unwrapped from their nucleosomes to allow the replication complex to do its work. The cell could use this period either to provide access to the DNA for transcription factors that would otherwise be excluded by the nucleosome, or to adjust the chromatin that assembles after replication in such a way as to facilitate or suppress transcription. The experi-ments outlined in the previous section have shown that bound components of the transcription initiation complex can resist chromatin assembly, making the proposed mechanism a feasible option.

   DNA replication is initiated only during S-phase of the cell cycle. During the rest of the time (i.e. the G1 and G2 phases and mitosis) the only DNA synthesis going on is that associated with the excision and replacement of damaged DNA by the DNA repair machinery. Initiation occurs at specific sites called origins of replication (OR). Quite what determines that a specific piece of DNA will act as an OR is not clear, although it is likely to involve both DNA sequence elements and DNA packaging. There are many thousands of OR in a typical cell, scattered through the genome at intervals of about, on average, 100 kb. Individual OR begin replication at different times during S-phase, with OR in transcriptional-ly-competent euchromatin tending to replicate early and those in heterochro-matin replicating late. However, it seems that individual OR are not regulated independently, but as groups termed replication units, with each unit usually containing between 20 and 80 individual origins. There is microscopical evidence that these units are physically clustered at specific sites in the nucleus (Fig. 7.11). All the origins in a unit start replication at the same time, but different units replicate at different times during S-phase. This results in regional variation in replication timing across the genome. This effect can be strikingly demonstrated by synchronizing cells growing in culture so they are all at more or less the same stage of the cell cycle, then labelling with DNA precursors, for a few minutes only (pulse-labelling), at different times in S-phase. This marks the DNA that happened to be replicating (i.e. incorporating new nucleotides) during the time the label is present. The labelled cells are allowed to proceed to the next metaphase, when chromosome spreads are prepared and treated in order to visualize the marked regions. These experiments soon revealed that blocks of noncoding centric heterochromatin replicated late in S-phase and that the AT-rich G-band regions tended to replicate later than the GC-rich R-band regions.

Progression of cells along defined differentiation pathways is dependent, among many other things, on DNA replication and cell division, with each cell cycle allowing the cell to reorganize patterns of gene expression that take it from one stage in the pathway to the next. These complex and coordinated changes are initiated by signals from neighbouring cells, by circulating hormones or by endogenous programmes. Some, at least, involve changes in chromatin structure. For example, the developmentally regulated silencing of homeotic genes in the *Drosophila* embryo is likely to involve the progressive spread of a suppressive chromatin conformation along the multigene complex. (See the section on the polycomb proteins in Chapter 10.) As cells change from one type to another

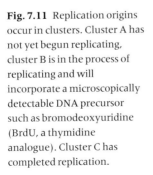

Fig. 7.11 Replication origins occur in clusters. Cluster A has not yet begun replicating, cluster B is in the process of replicating and will incorporate a microscopically detectable DNA precursor such as bromodeoxyuridine (BrdU, a thymidine analogue). Cluster C has completed replication.

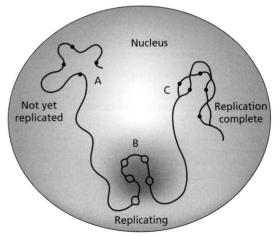

during differentiation and development, changes in gene expression and progression through the cell cycle occur together and appear to be inextricably linked. Cells can progress through many cell cycles without differentiating (stem cells and cultured cells stuck at a particular stage of differentiation do this all the time), but except in exceptional circumstances, non-cycling cells cannot differentiate. The molecular basis of this close connection is not clear. Indeed, it is remarkably difficult to find examples that show formally that the simple mechanism suggested by the *in vitro* nucleosome assembly experiments actually operates *in vivo*. Is it the simple dissociation of DNA from the nucleosome during replication that is important, or are more complex mechanisms involved? The question has been addressed in an *in vitro* system in which both transcription and DNA replication can be studied in synthetic nuclei.

It has already been noted that soluble (cytoplasmic) extracts from *Xenopus* oocytes can provide all the components necessary to assemble chromatin *in vitro*. If the extract is prepared so that it contains not only soluble components, but also membrane vesicles, it will build a membranous envelope around added DNA. In effect, it creates a replica nuclear envelope. Under these conditions, the DNA goes through one round of replication, with associated assembly of correctly spaced nucleosomes. These 'synthetic nuclei' provide an invaluable model system with which to study the relationship between DNA replication, chromatin assembly and transcription of specific genes. DNA constructs, containing the gene of interest, can be added to the egg extract, with or without potential transcription factors or regulatory proteins. Once the construct has been replicated and assembled into chromatin, its transcriptional activity can be tested.

This approach has been used to study expression of the developmentally regulated chicken β^A globin gene. Initial experiments were consistent with previous results with other genes, i.e. if the gene was assembled into chromatin in the absence of transcription factors, then it was inactive when tested for transcription. If the transcriptionally inactive, chromatin-packaged gene was taken through a round of DNA replication in the egg extract, then it remained inactive, showing that DNA replication itself does not automatically reactivate a previously inactive gene. However, if proteins extracted from chicken red blood cells (a source of transcription factors) were present *during* replication and chromatin assembly, then the gene was *active* when added to the transcription assay mix. These results are consistent with the idea that DNA replication provides an opportunity for transcription factors to gain access to DNA and reset the transcriptional potential of a gene (Fig. 7.12). They show, in addition, that DNA replication does not, itself, reset transcriptional potential. Importantly, replication also failed to *inactivate* the active, chromatin-packaged gene (i.e. the gene assembled into chromatin in the presence of RBC factors). It remains to be seen whether this 'memory' of transcriptional competence reflects the continued association of the RBC factors themselves or the persistence of a particular chromatin conformation through the DNA replication process.

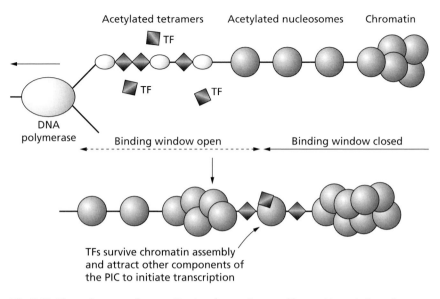

Fig. 7.12 The early stages of postreplication chromatin assembly provide a window of opportunity for transcription factor (TF) binding.

This is an important 'proof of principle', but the same results should not be expected with all genes or all transcription factors. Whether or not transcription factors are displaced during chromatin assembly will depend on their binding affinities relative to the core histones. Transcription factor binding affinities vary over at least two orders of magnitude.

Chromatin and the elongation stage of transcription

Nucleosomes can present a formidable barrier at the initiation stage of transcription. This is partly because the TBP binding to the TATA box is weak, compared with the binding of other transcription factors to their cognate sequences. Do nucleosomes also present a significant obstacle for RNA polymerase once it has left the initiation complex and begun moving along the DNA? The simple energetics of the process favour the polymerase. PolII itself is a large enzyme made up of about 12 subunits and with a molecular mass well in excess of 1 MDa. The transcribing enzyme remains associated with some general transcription factors (others are left at the promoter, Fig. 2.4), increasing its size still further. The polymerase complex dwarfs the nucleosome. (The molecular mass of the histone octamer is about 100 kDa.) The energetics of the process are also worth considering. Complete displacement of the histone octamer requires 10–50 kcal mol^{-1} (the exact value is very salt dependent). This energy cost becomes almost insignificant when we remember that every nucleotide in the growing RNA transcript requires the expenditure of one high-energy phosphate bond (about 10 kcal mol^{-1} *in vivo*). So, to transcribe the 150 or so base pairs of DNA associated with a nucleosome will require about 1500 kcal mol^{-1}. And this is just to assemble the RNA. The calculation takes no account of the kinetic energy involved in manoeuvring the macromolecules involved.

So, is the histone octamer irrelevant to the transcribing polymerase? The answer, at least *in vitro*, is clearly no. Although both PolII, and even much simpler, prokaryotic RNA polymerases can transcribe chromatin *in vitro*, the rate of transcription is very much lower than that which occurs on naked DNA. This results from a tendency for the enzyme to come to a standstill (to 'pause') when a nucleosome is encountered. The *in vitro* experiments are complex with technical difficulties that are even greater than those encountered in measuring protein–nucleosome interactions, but the message seems to be that polymerases on their own do not deal efficiently with chromatin. In order to enable them to do so *in vivo*, other factors must be involved. Likely candidates have recently been identified in the form of the protein family called 'elongins' and the chromatin remodelling complex 'FACT' (Chapter 8.)

Although *in vitro* transcription on chromatin templates is inefficient, it does occur to an experimentally useful extent. A series of experiments, primarily by Gary Felsenfeld and coworkers, have made the important point that simple (prokaryotic) polymerases can transcribe short pieces of DNA (about 200 bp) packaged into a nucleosome, without complete dissociation of the histone octamer. In other words, the polymerase manoeuvres around the octamer rather than completely displacing it. We would be entitled to ask how closely the mode of action of simple, prokaryotic polymerases is likely to resemble that of the PolII complex, but the finding that transcription without octamer displacement is *possible in principle* is an important one. This is particularly so because the octamer is unstable under the ionic conditions that prevail in the nucleus, unless it is protected by polyanions, such as DNA.

In order to transcribe the DNA coding strand, the polymerase unwinds about 30 bp of the DNA double helix. Unless the ends of the DNA are free to rotate, which will only be the case if we are dealing with short pieces of linear DNA, then the DNA must make topological adjustments in order to accommodate this unwinding. It does this by supercoiling. There is good experimental evidence that a wave of positive (R-handed) supercoiling precedes the transcribing polymerase and a wave of negative (L-handed) supercoiling follows it (Fig. 7.13a). The supercoiling is kept in check by topoisomerases, enzymes that release supercoils by making and repairing single- and double-stranded cuts in the DNA; essentially allowing the DNA helix to 'spin', in a controlled way, as if it were open ended. The wave of supercoiling will also favour the transfer of the histone octamer from the leading edge to the trailing edge of the polymerase. The reason is that the DNA wrapped around the nucleosome is negatively supercoiled. As core octamers are released from the DNA, then negative super-helical turns are released and can be used to cancel out the positive

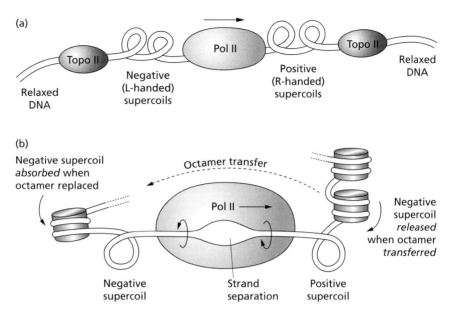

Fig. 7.13 DNA supercoiling ahead of and behind a transcribing polymerase is regulated by topoisomerases and may facilitate the transfer of histone octamers. (a) The strand separation that is part of the catalytic activity of the polymerase leads to positive supercoiling of DNA ahead of the polymerase and negative supercoiling of DNA behind. (b) Removal of a histone octamer ahead of the polymerase will release one negative supercoil and thereby cancel out one positive supercoil. Transfer of the octamer to the trailing edge of the polymerase complex and assembly of a nucleosome will cancel out one negative supercoil.

turns induced by the polymerase (Fig. 7.13b). In addition, if the octamer is transferred to the trailing edge of the polymerase, then it will encounter DNA that is already negatively supercoiled (i.e. 'bent' in the right way) and will readily reassemble (Fig. 7.13b).

The octamer transfer model makes energetic and topological sense and is consistent with much of the available experimental evidence. However, we have no idea how it is brought about. We do know that the octamer must be protected in some way at physiological ionic strengths, so cannot be allowed to wander far. Particularly relevant to octamer transfer models are experiments in which chromatin to be used for transcription is assembled with either native octamers, or octamers in which the histones have been chemically cross-linked to prevent dissociation. There was little difference in transcription efficiency between the two types of chromatin, indicating that disassembly of the nucleosome is not a necessary part of transcription on chromatin templates. This result argues against models that propose splitting of the nucleosome in order to allow the polymerase to pass through.

Further Reading

Transcription factor binding to nucleosomal DNA

Beato, M. & Eisfeld, K. (1997) Transcription factor access to chromatin. *Nucl. Acids Res.*, **25:** 3559–3563.

Felsenfeld, G. (1996) Chromatin unfolds. *Cell*, **86:** 13–19.

Owen-Hughes, T. & Workman, J. L. (1994) Experimental analysis of chromatin function in transcriptional control. *Crit. Rev. Euk. Gene Expr.*, **4:** 403–441.

Parajape, S. M., Kamakaka, R. T. & Kadonaga, J. T. (1994) Role of chromatin structure in the regulation of transcription by RNA polymerase II. *Ann. Rev. Biochem.*, **63:** 265–297.

Workman, J. L. (1998) Alteration of nucleosome structure as a mechanism of transcriptional regulation. *Ann. Rev. Biochem.*, **67:** 545–579.

Mouse Mammary Tumour Virus promoter

Eisfeld, K., Candau, R., Truss, M. & Beato, M. (1997) Binding of NF1 to the MMTV promoter in nucleosomes: influence of rotational phasing, translational positioning and histone H1. *Nucl. Acids Res.*, **25:** 3733–3742.

Smith, C. L. & Hager, G. L. (1997) Transcriptional regulation of mammalian genes *in vivo*. A tale of two templates. *J. Biol. Chem.*, **272:** 27493–27496.

Truss, M., Bartsch, J., Schelbert, A., Haché, R. J. G. & Beato, M. (1995) Hormone induces binding of receptors and transcription factors to a rearranged nucleosome on the MMTV promoter *in vivo*. *EMBO J.*, **14:** 1737–1751.

Transcription can be enhanced by a nucleosome

Elgin, S. C. R. (1988) The formation and function of DNase I hypersensitive sites in the process of gene activation. *J. Biol. Chem.*, **263:** 19259–19262.

Schild, C., Claret, F.-X., Wahli, W. & Wolffe, A. P. (1988) A nucleosome-dependent static loop potentiates oestrogen-regulated transcription from the *Xenopus* vitellogenin B1 promoter *in vitro*. *EMBO J.*, **12:** 423–433.

DNA replication and transcription

Almouzni, G. & Wolffe, A. P. (1994) Replication-coupled chromatin assembly is required for the repression of basal transcription *in vivo*. *Genes Dev.*, **7:** 2033–2047.

Almouzni, G., Méchali, M. & Wolffe, A. P. (1990) Competition between transcription complex assembly and chromatin assembly on replicating DNA. *EMBO J.*, **9:** 573–582.

Barton, M. C. & Emerson, B. M. (1994) Regulated expression of the β-globin gene locus in synthetic nuclei. *Genes Dev.*, **8:** 2453–2465.

Fangman, W. L. & Brewer, B. J. (1992) A question of time: replication origins of eukaryotic chromosomes. *Cell*, **71:** 363–366.

Chromatin and the elongation phase of transcription

Bednar, J., Studitsky, V. M., Grigoryev, S. A., Felsenfeld, G. & Woodcock, C. L. (1999) The nature of the nucleosomal barrier to transcription: direct observation of paused intermediates by electron cryomicroscopy. *Mol. Cell*, **4:** 377–386.

Shilatifard, A., Conaway, J. W. & Conaway, R. C. (1997) Mechanism and regulation of transcriptional elongation and termination by RNA polymerase II. *Curr. Opin. Genet. Dev.*, **7:** 199–204.

Studitsky, V. M., Kassavetis, G. A., Geiduschek, E. P. & Felsenfeld, G. (1997) Mechanism of transcription through the nucleosome by eukaryotic RNA polymerase. *Science*, **278:** 1960–1963.

Thoma, F. (1991) Structural changes in nucleosomes during transcription: strip, split or flip. *Trends Genet.*, **7:** 175–177.

Tsao, Y.-P., Wu, H.-Y. & Liu, L. F. (1989) Transcription-driven supercoiling of DNA: direct biochemical evidence from *in vitro* studies. *Cell*, **56:** 111–118.

Chapter 8: Chromatin Remodelling Machines

Introduction

There is now a wealth of experimental evidence to show that nucleosomes assembled over a promoter can block the initiation of transcription and that they do this by inhibiting the ability of transcription factors or other proteins to bind to their recognition sites on DNA. In some cases this repressive effect is helped along by mechanisms that encourage the positioning of a nucleosome directly over a crucial region of the promoter, thus preventing the possibility that, in some cells, these regions would fall in the linker DNA and thereby be accessible. But even with the help of such positioning effects, the presence of nucleosomes at and around a promoter is not, in all cases, enough to ensure that the promoter is completely inactive. As I have already discussed, some transcription factors are able to bind very well to their cognate DNA, even when it is constrained within a nucleosome, and with others the inhibition of binding, although severe, may not be enough to ensure, in itself, the complete suppression of transcription. To deal with this, enzyme-based mechanisms have evolved that either amplify or suppress the inherently repressive effects of chromatin. These mechanisms will be considered in two sections. The first will deal with enzymes that directly manipulate nucleosome structure in an ATP-dependent manner and that render the DNA more accessible to transcription factors and DNA-binding proteins. The second deals with the enzymes that alter chromatin by regulating the level of acetylation of the core histone N-terminal domains, namely the histone acetyltransferases and deacetylases.

Nucleosome remodelling enzymes

Enzyme complexes have evolved whose job is to displace or disrupt nucleosomes and thereby increase the accessibility of selected regions of DNA. As has often been the case, the first of these complexes was identified through genetic studies in yeast.

The SWI/SNF family

The story begins with studies on two types of yeast mutants, one with a reduced ability to grow in medium containing sucrose (sucrose non-fermenting, SNF, mutants) and the other with aberrant mating type switching (SWItching mating type, SWI, mutants). In some cases, the same mutation was found to cause both phenotypes. The genes responsible for each of these two different phenotypes are now referred to, collectively, as SWI/SNF ('switch-sniff') genes. The

SWI/SNF genes do not, themselves, play a direct role in either sucrose metabolism or mating type switching, but are instead required for the proper expression of a small subset of genes whose products *are* involved in these cellular functions.

A clue that the regulation of this subset of genes operated through chromatin came from the finding that additional mutations that tended to destabilize or disrupt normal chromatin structure prevented the phenotypic effects of the SWI/SNF mutants. In genetic terms, they acted as *suppressor* mutations. These suppressor mutations could be subtle (e.g. a point mutation of a core histone) or not so subtle (e.g. a 50% reduction in the level of histones H2A and H2B). Thus, mutations that cause an abnormal nucleosome structure or simply reduce nucleosome density render the SWI/SNF gene products unnecessary (Fig. 8.1). This interpretation of the genetic data was confirmed when a large protein complex containing SWI/SNF gene products was isolated from yeast by conventional biochemical methods and shown to have the ability to remodel nucleosomes *in vitro*, i.e. to change the relationship between histones and DNA such that nuclease sensitivity was measurably altered. The SWI/SNF complex can be assayed *in vitro* by using positioned nucleosomes as substrate and measuring changes in the characteristic pattern of DNaseI cleavage. As described in Chapter 3, DNaseI cuts preferentially at points where the minor groove is fur-

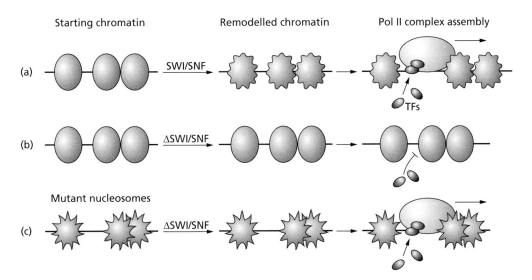

Fig. 8.1 Mutations to the SWI/SNF remodelling complex inhibit transcription from some genes but can be complemented by mutations that alter nucleosome structure. (a) In normal (wild type) cells SWI/SNF remodelling of nucleosome positioned over the promoters of certain genes is required for assembly of the transcription initiation complex. (b) Mutations to SWI/SNF that prevent remodelling will inhibit transcription from these genes. (c) A second mutation that leads to an altered nucleosome structure (e.g. a core histone mutation) can render SWI/SNF remodelling unnecessary and restore transcription to near-normal levels.

thest from the histone core, i.e. at approximately 10 bp intervals (Fig. 3.5). SWI/SNF converts this nucleosomal cleavage pattern to one more closely resembling that of free DNA (Fig. 8.2; note that as the concentration of SWI/SNF increases, the original preferential cutting sites become less prominent and new cutting sites appear that correspond to those present in free DNA). The energy necessary for this nucleosome remodelling comes from the cleavage of ATP, and none of the changes shown in Fig. 8.2 are seen in the absence of ATP. Cleavage of ATP is catalysed by the SWI2/SNF2 subunit (Table 8.1). The ATPase domain of SWI2/SNF2 resembles that present in helicases, enzymes that can use the energy from ATP cleavage to unwind DNA (or RNA) duplexes.

The mechanism by which the SWI/SNF complexes remodel chromatin remains obscure, although studies of the properties of remodelled nucleosomes have provided some useful clues. Most nucleosomal DNA remains associated

Fig. 8.2 SWI/SNF remodelling changes the nuclease digestion pattern of nucleosomal DNA so that it more closely resembles that of free DNA. The DNA fragments generated by nuclease digestion of chromatin and separated by electrophoresis are shown in the left-most track. Pretreatment of chromatin with increasing amounts of human SWI/SNF causes the progressive appearance of additional DNA fragments that are normally found only if the same DNA sequence is dissociated from histones and digested as free DNA (right-most tracks). SWI/SNF has no effect on the nuclease digestion properties of free DNA (right-hand tracks, −and+SWI/SNF). (Data from Imbalzano *et al.* 1994, *Nature* **370**, 481–485.)

Table 8.1 Chromatin remodelling complexes

Complex	Organism	Subunits	ATPase	Notes
SWI/SNF family				
SWI/SNF	*S. cerevisiae*	11	SWI2/SNF2	1
RSC	*S. cerevisiae*	15	STH1	2
hSWI/SNF	*H. sapiens*	~10	hbrm, BRG-1	3
Brahma	*D. melanogaster*	>7	BRM	4
Mi-2 family				
Mi-2	*X. laevis*	6	Mi-2	5
NuRD	*H. sapiens*	>7	Mi-2	5
ISWI family				
ACF	*D. melanogaster*	2	ISWI	6
CHRAC	*D. melanogaster*	5	ISWI	7
NURF	*D. melanogaster*	4	ISWI	8

1. Very low abundance, possibly less than 200 per cell. Mutants are viable. ATPase is stimulated by ssDNA, dsDNA and nucleosomes. Remodelling requires a 1:1 ratio of complex to nucleosomes and results in a nuclease digestion pattern similar to that of naked DNA.
2. At least three subunits are homologous to those in SWI/SNF; has similar enzymatic properties and similar effects on chromatin. At least 10× more abundant than SWI/SNF and loss of function mutants are nonviable.
3. The hbrm and BRG-1 ATPases are in different, but related, complexes. Both contain an SNF5 homologue (hSNF5).
4. The BRM protein is involved in the transcriptional activation of homeotic genes during development. -/- mutants are non-viable. Complex also contains SNR1, an SNF5 homologue.
5. The Mi-2 and NuRD complexes are closely related. Mi-2 was first identified as an autoantigen in the human disease dermatomyositis. Other subunits include the protein MTA2, so named because it is closely related to a protein identified in many metastasizing human cancers and labelled metastasis associated antigen 1 (MTA1). These complexes also contain the histone deacetylases HDAC1 and HDAC2 and proteins that bind methylated DNA.
6. Together with a histone chaperone such as NAP-1 or CAF-1, will assemble chromatin in periodic arrays. Also facilitates TF binding and modulates internucleosomal spacing.
7. Gives regular spacing of nucleosomes in chromatin and increased restriction enzyme accessibility.
8. ATPase stimulated by nucleosomes but not by DNA alone. Makes DNA in chromatin more accessible to cutting by micrococcal nuclease, but not to such an extent that it resembles naked DNA. Active at sub-stoichiometric levels.

with the histone octamer, even though rotational phasing is randomized and the DNA becomes more accessible to DNA-binding proteins. Perhaps surprisingly, the DNA double helix is not unwound. There is strong evidence that the core histone tails are needed for SWI/SNF remodelling and that acetylation of the tails reduces remodelling activity. Displacement, splitting or major rearrangment of the H2A/H2B dimers is not required, nor is the H3/H4 tetramer split. Nonetheless, the histone octamer is more easily displaced in remodelled chromatin and the length of DNA associated with the histone core is reduced.

The SWI/SNF complex is present in small amounts in yeast cells – possibly as few as 200 molecules per cell – and genetic evidence indicates that it is involved in the regulation of a limited number of genes. This has been confirmed by the use of DNA arrays (specifically high-density oligonucleotide arrays, HDAs), to measure simultaneously the levels of large numbers of mRNA transcripts in yeast mutants lacking one or another of various proteins involved in gene regulation. (The procedure is outlined in Box 4.) In mutants lacking the Swi2/Snf2 ATPase (and therefore without a functional SWI/SNF remodelling complex), mRNA transcripts from only 126 genes, of 5695 tested, were reduced more than twofold. Surprisingly, transcripts from a larger number of genes (203) were more than doubled, suggesting that, in some cases, the SWI/SNF complex is involved in gene *repression*. (It is important to remember that not all the genes showing altered expression will necessarily be *directly* regulated by SWI/SNF. Some may be influenced instead by the *products* of SWI/SNF-dependent genes.) Perhaps the message to take from these results is that remodelling chromatin to improve the accessibility of its DNA can be used to facilitate the binding of both repressive and activating proteins. It is interesting that recent experiments have shown that the SWI/SNF complex is particularly important for the regulation of genes that are expressed late in mitosis, when the genome is still partially condensed.

A second yeast complex, termed RSC (remodel the structure of chromatin), contains proteins closely homologous to components of the SWI/SNF complex and remodels chromatin in a very similar way. But RSC is present in at least 10-fold greater amounts than SWI/SNF and is essential for viability (Table 8.1). In addition, homologues of the SWI/SNF complex have now been found in both mammals and *Drosophila* (Table 8.1). Like SWI/SNF and RSC, these complexes contain a DNA-dependent ATPase activity. In *Drosophila*, the ATPase is the product of the *Brahma* (*brm*) gene, whose protein product is responsible for the maintenance of the transcriptional competence of selected genes during development (see Chapter 11). The human complex is called, unimaginatively, hSWI/SNF. There are two human ATPases that are homologues of the *Drosophila Brahma* protein, namely hBRM and BRG-1.

Mammalian SWI/SNF complexes

The role of chromatin remodelling complexes in mammals is well illustrated by transcription of the β-globin gene under the control of the EKLF transcription factor. EKLF (erythroid Krüppel-like factor) is essential for the initiation of transcription from the β-globin locus early in embryonic development. Its action can be studied *in vitro* using the β-globin gene assembled into chromatin (see Chapter 7). Such experiments soon revealed that EKLF alone was not sufficient for optimum transcription; a second factor, present in crude nuclear extracts, was also necessary. This factor was purified from cultured erythroleukaemia cells using its ability to enhance β-globin transcription as an

Box 4 DNA microarrays

A DNA array can be constructed from synthetic oligonucleotides (about 25bp) or from DNA fragments (more than 100 bp) made by PCR. The PCR products can be from cDNA libraries (for expression analysis) or from genomic DNA, using primers designed to scan across a defined (sequenced) region of the genome.

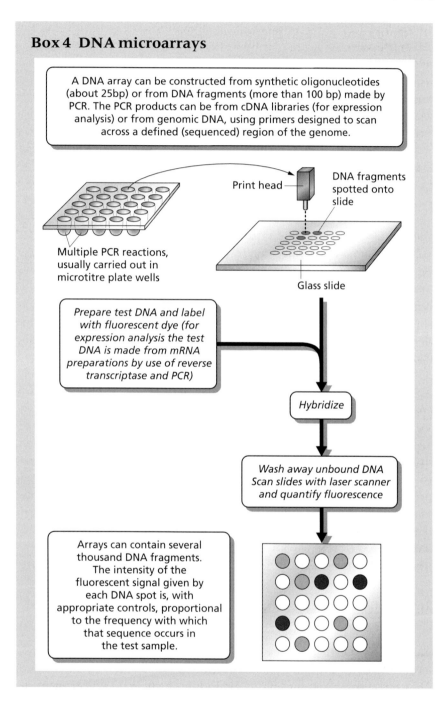

Print head

DNA fragments spotted onto slide

Multiple PCR reactions, usually carried out in microtitre plate wells

Glass slide

Prepare test DNA and label with fluorescent dye (for expression analysis the test DNA is made from mRNA preparations by use of reverse transcriptase and PCR)

Hybridize

Wash away unbound DNA Scan slides with laser scanner and quantify fluorescence

Arrays can contain several thousand DNA fragments. The intensity of the fluorescent signal given by each DNA spot is, with appropriate controls, proportional to the frequency with which that sequence occurs in the test sample.

assay, and was shown to be a large, multiprotein complex containing several proteins (including BRG-1) that are homologous to components of the yeast SWI/SNF complex. The complex, named E-RC1, can remodel positioned nucleosomes (as shown in Figs 8.1 and 8.2) and induces a DNaseI hypersensitive site upstream of the β-globin promoter in essentially the same position as the one that appears when β-globin is switched on *in vivo*. Interestingly, E-RC1 is needed for optimum transcription even when EKLF is allowed to bind to the promoter before chromatin assembly, so it must act *after* transcription factor binding. E-RC1 does not enhance transcription of free β-globin DNA, nor does it enhance transcription of all genes assembled as chromatin. It is clearly not a totally nonspecific activator of transcription. At least one of the proteins present in E-RC1, and, presumably, in other human SWI/SNF complexes, must be responsible for targeting the complexes to specific promoters. This may be brought about by binding to transcription factors or to other promoter specific markers.

The potential involvement of chromatin remodelling systems in a variety of nuclear functions is illustrated by the human complexes containing the ATPases hbrm or BRG-1. Both hbrm and BRG-1 bind to the protein product p105Rb of the RB tumour-suppressor gene and increase its ability to slow cell growth. This is done by *reducing* transcription of genes regulated by the transcription factor E2F and necessary for progression through the cell cycle. This is discussed in more detail later. The important point for now is that at least some components of mammalian SWI/SNF are involved in *suppression* of transcription rather than its activation, exactly as predicted by the results with yeast mutants cited earlier.

The conclusion that mammalian SWI/SNF components are more than just transcriptional activators is born out by the phenotypes of knockout mice lacking hbrm or BRG-1. Despite the fact that *Drosophila* Brahma is involved in the regulation of genes that play a crucial role in development, mice that lack both copies of *brm* do not show major developmental defects. However, they are 10–15% bigger than normal, presumably a consequence of defective cell growth control. Fibroblasts from such mice show impaired growth arrest at confluence. In contrast, lack of the *BRG-1* gene causes early death *in utero*. The sometimes unexpected phenotypic effects of mutations in the protein components of chromatin remodelling systems are consistent with their involvement in a network of protein–protein interactions with multiple, sometimes redundant functions.

The ISWI family

Other chromatin remodelling complexes, more distantly related to SWI/SNF, have been identified in *Drosophila* nuclear extracts. In fact, the NURF complex (Table 8.1) was the first chromatin remodelling complex to be isolated and biochemically characterized, using an assay based on changes in the micrococcal nuclease digestion properties of nucleosome arrays. The assay is subtly different from that used for SWI/SNF in Fig. 8.2 in that it measures changes in nucleosome *spacing* rather than changes in individual nucleosome core particles. The extraction and assay procedures developed by Carl Wu and colleagues at N.I.H.

for work on the NURF complex have since been used to identify two other remodelling complexes in *D. melanogaster*. These are listed in Table 8.1. The three are biochemically and functionally different, although all of them have the same DNA-dependent ATPase/helicase subunit, a protein termed ISWI (Imitation SWItch). ISWI resembles the yeast protein SWI2/SNF2 in its ATPase domain, although the ATPase domain of ISWI cannot substitute for that of SWI2 in 'domain-swap' experiments. The important property shared by all three complexes is that they alter the positioning of nucleosomes along DNA, even though they do so in rather different ways, at least in *in vitro* assays. In this respect they differ from SWI/SNF, which seems to be primarily a remodeller of the nucleosome itself, rather than a rearranger of nucleosome arrays. (However, a *consequence* of nucleosome remodelling by SWI/SNF might well be chromatin rearrangement through the subsequent action of other factors.) The existence of these different complexes suggests that eukaryotic cells may contain different chromatin remodelling activities targeted to different regions and involved in different nuclear processes. This is not unexpected, given that chromatin remodelling is an integral part not only of transcription initiation and elongation, but also of DNA replication, DNA repair and, possibly, cell cycle progression.

The Mi-2 family

The remodelling complexes based on the Mi-2 ATPase have two features that make them stand out from the other remodelling activities identified thus far. First, they have an inbuilt targeting mechanism. One of their protein subunits binds preferentially to methylated DNA. Increased levels of methyl-cytosine are associated with chromatin condensation and genetic silencing, as discussed in Chapter 10. Second, Mi-2 complexes include the histone deacetylases HDAC1 and HDAC2. Histone underacetylation is also a feature of condensed, transcriptionally inactive chromatin. The histone-deacetylating enzymes are discussed later in this chapter and we will return to the Mi-2/NuRD complex later in the book; but note for now that the Mi-2 complex introduces the important point that these different chromatin-modifying activities do not operate independently *in vivo*. They act instead in a coordinated manner, even when not physically associated, and the action of one may amplify or suppress the activity of another.

Histone acetyltransferases (HATs)

All histones in the nucleosome core particle are subject to various post-translational modifications, most of which are made to the N-terminal domains located on the nucleosome surface. These modifications have been described in Chapter 4. The most frequent of these modifications is the acetylation of specific lysine residues. Acetylation is a dynamic process and histone acetate groups turn over with half-lives ranging from a few minutes to several hours. Turnover results from the action of two families of enzymes, the histone acetyltransferases (HATs), catalysing the transfer of an acetate group from acetyl CoA to a lysine ε-amino group, and the histone deacetylases (HDACs), catalysing the removal of these acetates (Fig. 3.14). Attachment of an acetate group converts the lysine side-chain from positively charged to neutral, a change that will inevitably reduce the avidity with which it binds to DNA. Although the N-terminal tail domains, acetylated or not, seem to have only slight effects on the structure of the nucleosome core particle itself, they do have demonstrable effects on higher-order chromatin structure (Chapter 5). They also influence the ability of chromatin to bind transcription factors and other proteins *in vitro* (Chapter 7). Genetic experiments in yeast (described in Chapter 10) have shown that the histone N-terminal tails are involved both in controlling expression of certain genes and in cell cycle progression. So, histone acetylation qualifies as a modifier of chromatin structure and function. A better appreciation of its crucial importance as a regulator of transcription has come from recent work on the enzymes responsible for the acetylation/deacetylation cycle, the HATs and HDACs.

A cytoplasmic HAT

Early biochemical work identified two classes of histone acetyltransferase distinguishable by their intracellular location. The cytoplasmic enzymes were classified as members of the HAT B family and the nuclear enzymes as HAT A. The cytoplasmic enzymes are responsible for the post-translational acetylation of newly synthesized H4 at lysines 5 and 12, a modification that is important for chromatin assembly on newly replicated DNA. The nuclear enzymes are expected to act on preformed chromatin. The first HAT whose gene was cloned and sequenced was a cytoplasmic enzyme, designated Hat1, from the yeast *Saccharomyces cerevisiae*. It was pleasing to find that when the gene was cloned into bacteria and expressed, the recombinant protein could acetylate free H4 at lysines 12 and 5, exactly as one would expect. However, the story was very soon complicated by the discovery that in the yeast cell, Hat1 was complexed with a second protein, designated Hat2. This protein, despite its name, does not itself have histone acetyltransferase activity, but it does seem to play a role in determining the substrate specificity of Hat1. The complex containing the two proteins could still acetylate free H4 *in vitro*, but only at lysine 12. This finding makes the important point that complexing HATs with other proteins can change their substrate specificities. Mutations in the *hat1* gene led to a reduction in overall HAT activity, but had no other obvious phenotypic effects. This could be because other HATs can take over cytoplasmic H4 acetylation if Hat1 is missing. [It should be noted, in passing, that the lack of a detectable growth phenotype in *hat1*− strains is consistent with the fact that yeast in which H4 lysines 5 and 12 have been substituted with arginine grow perfectly well in the laboratory (see Chapter 10). This tells us that acetylation of lysines 5 and 12 is not *essential* for chromatin assembly, although the fact that this precise modification of newly synthesized H4 is found in organisms as diverse as yeast and humans suggests that it does confer some selective advantage.]

Nuclear HATs and transcriptional coactivators

The clearest evidence linking HATs to transcription initiation came with the purification of a HAT B (designated p55) from the ciliated protozoan *Tetrahymena thermophila* and the cloning and sequencing of the gene encoding it. The gene was found to be closely homologous to yeast *GCN5*, a gene that was known, through genetic studies, to encode a regulator of transcription. Subsequent studies on the protein product of yeast *GCN5* (i.e. Gcn5) showed that, like p55, it had HAT activity. Close homologues of yeast Gcn5 have since been found in various species, including humans. Some examples are listed in Table 8.2.

The Gcn5-related enzymes are not the only nuclear HATs. The link between HATs and transcription was further strengthened by the discovery that other regulators of transcription, unrelated to Gcn5, also had HAT activity. These include p300 and CBP, two very similar proteins that are involved, amongst other things, in cell cycle progression and the activation of genes in response to the cell-signalling molecule cyclic AMP (cAMP). CBP and p300 are, for most purposes, functionally equivalent and are often referred to jointly as CBP/p300. They are members of a class of proteins known as transcriptional 'coactivators'. These proteins do not themselves show any sequence-specific DNA binding, they are not transcription factors, but they can associate, more or less specifically, with sequence-specific DNA-binding proteins, which serve to target them to specific genes or genomic regions. Figure 8.3 shows some of the transcription factors that bind to CBP/p300 and the regions of the protein to which they bind.

Table 8.2 Some histone acetyltransferase (HAT) complexes in yeast and humans

HAT	Complex	Size	Other protein components
Yeast (*S. cerevisiae*)			
Gcn5	HAT-A2	~200 kDa	Ada2,3 + others
	ADA	~900 kDa	Ada2,3 + others
	SAGA	~2 MDa	Ada1–5, Spt and TAF$_{II}$ proteins, Tra1
Esa1	NuA4	1.3 MDa	Tra1 + various others
Human			
TAF$_{II}$250*	TFIID		TBP + various other TAF$_{II}$ proteins
CBP/p300	various	various	Gene-specific activators PCAF, ACTR, SRC-1; all of which are also HATs
PCAF, hGcn5	not named	~20 proteins	Counterparts of Ada, Spt and Tra proteins + TAF$_{II}$s + others

*The equivalent TAFs in yeast and *Drosophila* also have HAT activity.

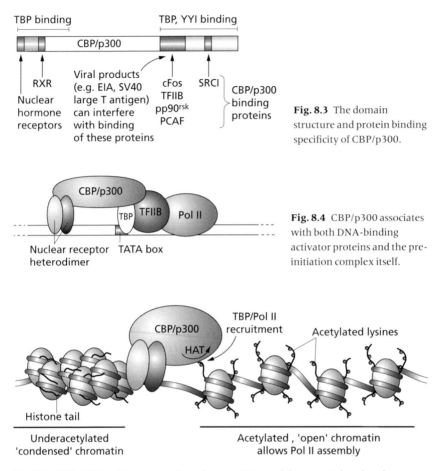

Fig. 8.3 The domain structure and protein binding specificity of CBP/p300.

Fig. 8.4 CBP/p300 associates with both DNA-binding activator proteins and the pre-initiation complex itself.

Fig. 8.5 CBP/p300 has histone acetyltransferase activity and the potential to alter the structure of chromatin at promoters.

CBP/p300 is a huge protein of more than 2400 amino acids, so is well able to play a multifunctional role involving interactions with several other proteins. Domains at the N-terminal and C-terminal ends of CBP/p300 are able to bind to the transcription initiation complex via TBP, although the mechanism by which this binding could bring about transcriptional activation remains something of a mystery (Fig. 8.4).

The finding that CBP/p300 has HAT activity led to the appealing model illustrated in Fig. 8.5. The recruitment of these CBP/p300 to DNA sequences packaged as chromatin results in increased histone acetylation and an opening up of chromatin, possibly as a result of weakened internucleosomal bridges or through disruption of complexes between the histone tails and nonhistone proteins mediating chromatin compaction. Improved access for components of the transcription initiation complex leads to an increase in transcription.

At least three of the proteins that can bind to CBP/p300 (namely PCAF, SRC-

1 and ACTR) also have HAT activity. The reason for having three HATs in the same complex is not entirely clear, although it is known that different activators require different HATs. For example, activation of transcription by the cAMP response element binding (CREB) protein requires the HAT activity of CBP, whereas activation by the retinoic acid receptor (RAR) requires that of PCAF. Although the HAT activity of CBP is not required for activation via the RAR, the protein itself is, presumably because it is necessary for correct assembly of the coactivator complex. This illustrates the important point that the HATs may have functions beyond their catalytic activities. It is reasonable to suppose that the different HAT catalytic subunits have different specificities, perhaps acetylating different histones, or even nonhistone proteins (see below). Different patterns of acetylation may be necessary for optimal activation of different sets of genes. It is also possible that the same gene may require different HATs for optimal activation depending on the stage of development or differentiation. There is no reason to assume that a particular gene will always require the same HAT activity for activation.

In addition to multiple HATs, CBP/p300 can recruit a protein kinase, $pp90^{rsk}$ (Fig. 8.3). This protein is required for activation of genes that respond to Ras, one of the many GTP-binding proteins that relay signals from cell surface receptors to the nucleus. This finding emphasizes the enormous potential of the large complexes assembled around CBP/p300 and other coactivator proteins as mediators of intracellular signals. These complexes may be the sites at which the protein kinase cascades that transmit signals from cytoplasm to nucleus actually impact on the genome. It has already been shown that the HAT activity of CBP/p300 can be altered by changing its phosphorylation state and it is a reasonably safe prediction that other chromatin modifying enzymes will be found to behave in the same way.

It is important to remember that there are several histone modifications in addition to acetylation. These have been discussed in Chapter 4. Enzymes responsible for phosphorylation and methylation of histones (at specific serine and lysine residues, respectively) have recently been described and others will no doubt follow. It is likely that modifications will be found to be complementary or antagonistic. For example, an H3 peptide phosphorylated at serine 10 is a better *in vitro* substrate for the yeast HAT Gcn5 than the equivalent unphosphorylated peptide. It would be very surprising if there were not functionally significant 'cross talk' between different histone modifications *in vivo*. Working out the significance of the many possible combinations of histone modifications is likely to be a major preoccupation in the chromatin field over the next few years.

Larger and larger complexes

Attempts to purify nuclear HATs by conventional biochemical means (such as column chromatography) have shown that many nuclear HATs in addition to

CBP/p300 exist in the form of large, multi-subunit complexes, often with 20 or more protein components and molecular weights in excess of 2 MDa (2×10^3 kDa). Characterization of such complexes is something of a technical challenge, particularly in view of the likely variability noted earlier and the ever-present possibility that proteins are lost or gained during the purification process. Some of the complexes described so far are listed in Table 8.2, together with some of the different members of the HAT enzyme family so far identified. Note that at least one component of the basal transcription initiation complex ($TAF_{II}145$ in yeast, $TAF_{II}250$ in humans) has HAT activity. The significance of this is still unclear, but could be related to the recent observation that other protein components of the complex are acetylated *in vivo*, albeit at a relatively low level compared with the histones. It may be that, although the core histones provide useful substrates with which to assay HAT activity *in vitro*, the real *in vivo* substrates for some HATs may be nonhistone proteins.

The oligonucleotide arrays used to measure levels of large numbers of individual mRNAs in wild type and mutant yeast have been used to test the effect of loss-of-function mutations of *GCN5*, the gene encoding the catalytic component of the SAGA complex, and of $TAF_{II}145$, a HAT component of the yeast basal transcription initiation complex. Loss of *GCN5* significantly changed the transcription of only 5% of genes tested, comparable with the effect of loss of SWI/SNF function, whereas loss of $TAF_{II}145$ changed 16%. By comparison, nearly 70% of genes were found to be dependent upon $TAF_{II}17$, a small protein present in both the SAGA and basal transcription initiation complexes. Clearly these figures will be influenced by the extent to which other mechanisms can take over from those eliminated by the mutations being tested. But, even allowing for this, they do show how particular complexes, and even individual components within them, may be required for the transcription of only a small proportion of genes. It will be particularly interesting to see whether the genes dependent on particular chromatin modifying activities have anything in common. It has already been noted that among the genes dependent on $TAF_{II}145$, a large number are essential for cell cycle progression. Yeast strains carrying a mutation in the *GCN5* HAT gene also have a slow growth phenotype.

Histone deacetylases

The histone deacetylases and acetyltransferases are complementary enzymes that, between them, maintain the level of histone acetylation throughout the genome, and possibly in the cytoplasm as well. It is convenient to consider them under separate headings, but it is important to remember that they are very much part of the same system of gene regulation. Like the HATs, the HDACs are often part of multisubunit complexes and some proteins are found in both types of complex.

The characterization of the HDACs and cloning of the genes that encode them has a similar history to that of the HATs. The first HDAC to be purified to near homogeneity was isolated from human cultured cells by a clever procedure involving affinity chromatography on a matrix containing the specific HDAC inhibitor trapoxin. Two proteins were retained by the affinity matrix. Microsequencing of these proteins showed that one was RbAp48, a protein that associates with the product of the RB tumour suppressor gene, while the other, now known as HDAC1, was a novel protein containing the trapoxin-binding site and the deacetylase catalytic domain. Screening of DNA sequence databases soon identified expressed human DNAs that encoded HDAC peptides, allowing isolation and sequencing of full-length DNA clones. Evidence for the involvement of histone deacetylases in gene regulation came with the finding that the predicted amino acid sequence of human HDAC1 was 60% identical to that of the protein encoded by the yeast gene *RPD3*, a known transcriptional regulator. The yeast Rpd3 protein was very quickly shown also to be a histone deacetylase. At about the same time, biochemical experiments with the yeast *S. cerevisiae* succeeded in isolating two protein complexes with histone deacetylase activity, one of which contained Rpd3 and the other a novel deacetylase named HDA1. Three homologues of HDA1, designated HOS1, HOS2 and HOS3 were identified quickly, making it apparent that, even in a simple eukaryote such as *S. cerevisiae*, the histone deacetylases are a diverse family of enzymes. The situation may become even more complex, because not all deacetylases need be homologous to Rpd3 or HDA1. The protein encoded by the yeast silencing regulator gene *SIR2* (see Chapter 10) has been shown to have histone deacetylase activity, but is not homologous to known deacetylases in yeast or other organisms. The deacetylase activity of Sir2 was not detected earlier because this activity is entirely dependent on the presence of mM concentrations of NAD. This unusual property gives Sir2 the potential to modify chromatin structure in a way that is sensitive to changes in intermediary metabolism.

Like the HATs, the HDACs often function as part of multi-subunit complexes involving a large, and ever-increasing, number of other proteins. Whether this is the case for all the various members of the HDAC family remains to be seen, but it is certainly true for HDAC1 and HDAC2 in higher eukaryotes. For these enzymes, there seems to be a core catalytic complex consisting of HDAC1 and/or HDAC2 (they are usually found together) and RbAp48 and the closely related protein RbAp46 (Fig. 8.6). These two proteins are also found in association with the cell cycle regulating retinoblastoma protein, Rb. This core complex is found associated with a bewildering variety of other proteins, many of which have known or suspected functions in gene regulation. Fortunately, for the purposes of describing what is going on, most of these interactions are mediated by one or the other of two key proteins, namely Sin3a and Mi-2. The situation is outlined in Fig. 8.6. Sin3 complexes contain, in addition to HDAC1/2 and RbAp46/48, two smaller proteins SAP30 and SAP18 (SAP=Sin3 associated

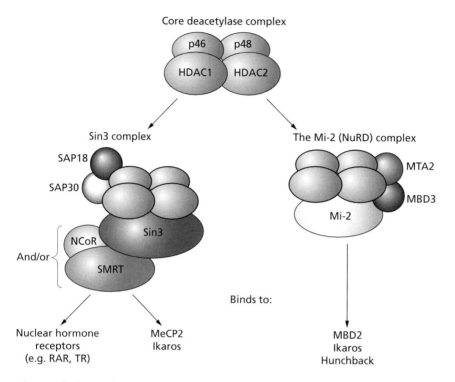

Fig. 8.6 The histone deacetylases HDAC1 and HDAC2 function as part of multiprotein complexes. To be catalytically active, HDAC1 and HDAC2 must be complexed with other proteins, at least with p46 and p48, two proteins that also associate with the retinoblastoma protein Rb. This catalytically active core complex can then associate with other proteins to form larger complexes of two general types, the Sin3 complex and the Mi2 complex. The latter has chromatin remodelling activity. These complexes are targeted to specific genes and chromosome regions by association with additional proteins, such as nuclear hormone receptors and the methyl-CpG-binding proteins MeCP2 and MBD2.

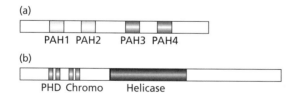

Fig. 8.7 The domain structures of mammalian Sin3 and Mi-2. The Sin3 protein contains four paired amphipathic helix (PAH) domains that probably mediate interactions with other proteins and help assemble the deacetylase complex. (An amphipathic helix is one in which amino acids of opposite charges are located on opposite sides of the helix; they are often involved in protein–protein binding). Mi-2 is the largest component of the Mi-2 deacetylase complex (also known as the NuRD complex). It contains two plant homeodomains (PHD) two chromodomains and a distinctive SWI/SNF helicase/ATPase domain that is necessary for chromatin remodelling.

protein). Sin3 contains four paired amphipathic helix (PAH) domains that are involved in interactions with the other proteins in the complex (Fig. 8.7a). The Sin3 complex can bind two other relatively large proteins, NCoR (Nuclear hormone receptor CoRepressor) and SMRT (Silencing Mediator of Retinoid and Thyroid hormone receptor) that can, as their names suggest, mediate binding to receptors for hormones such as oestrogen, retinoic acid or vitamin D. Alternatively, the Sin3 complex can associate with DNA-binding heterodimers such as mad/max or to the methyl-DNA-binding protein MeCP2 (see Fig. 8.6). These associated proteins all have the ability to target the Sin3 deacetylase complex to particular parts of the genome.

The other group of HDAC1/2 complexes is based on the protein Mi-2. The domain structure of Mi-2 is shown in Fig. 8.7b. Its most striking feature is a distinctive helicase/ATPase domain homologous to that found in the helicase subunit of the SWI/SNF complex. The Mi-2 protein has ATPase activity and, as noted earlier, the complex has nucleosome remodelling activity. (It is also known as the NuRD complex, for Nucleosome Remodelling histone Deacetylase.) In addition to HDAC1/2, RbAp46/48 and Mi-2, the complex contains a methyl-DNA-binding domain (MBD) protein MBD3 and MTA2, a protein homologous to the metastasis associated protein MTA1 found in human tumours. As indicated in Fig. 8.6, the Mi-2/NuRD complex can associate with various other proteins, at least some of which can target it to specific regions of the genome.

There is a long list of proteins that associate with either the Sin3 or Mi-2/NuRD complexes, only some of which are listed in Fig. 8.6. Proteins occasionally associate with both. Many are DNA-binding proteins and their primary function is to target the HDAC complexes to specific regions of the genome. For example, the transcription factor heterodimer mad/max targets the complex to growth-regulating genes as part of the mechanism that silences them at appropriate times in differentiation. The retinoic acid receptor (RAR) plays a similar role for genes that respond to retinoic acid, as discussed later. Proteins such as NCoR and SMRT have no DNA-binding ability themselves, but function to link the core Sin3 complex to proteins that do.

The situation may seem rather complicated, and the number of proteins involved (with their bizarre acronyms) is certainly large, but there are some simplifying generalities that may help. The proteins that make up the deacetylase complexes can be grouped into three, *nonexclusive*, functional categories:

1 *Catalytic proteins*. HDAC1 and HDAC2 are the only components with deacetylase catalytic sites, but the two proteins RbAp46 and RbAp48 may have an important role in facilitating catalytic activity. HDAC1, in recombinant form, expressed from *Escherichia coli* is catalytically inactive and the smallest active complex that can be isolated biochemically contains RbAp48. It may be that some proteins in the complexes can influence the substrate specificity of the two catalytic subunits, although this remains to be proved. The protein Mi-2 (an ATPase) also falls into this category.

2 *Targeting proteins.* Proteins such as mad/max and RAR/RXR heterodimers target the complexes to specific genes while MeCP2 targets methylated DNA. (The physical interactions between HDACs and methyl DNA-binding proteins mediate the close link between DNA methylation and histone deacetylation. This important topic is dealt with in more detail in Chapter 10. It is also interesting to note that the protein RbAp48 is associated with the CAF1 chromatin assembly complex. It may be involved in the assembly of complexes that deacetylate newly assembled chromatin.

3 *Mediating proteins.* The HDACs themselves are far too small to be able to bind to all the various targeting proteins on their own. Only by using large, intermediate proteins such as Sin3, Mi-2 and NCoR can such a large range of partners be accommodated.

The same sort of analysis can be applied to the HAT complexes, although in this case one protein, CBP/p300, serves as both an enzyme and a large mediator.

The nuclear receptors

The proteins collectively known as *nuclear receptors* have the ability to bind to specific hormones and activate appropriate sets of genes in a hormone-sensitive fashion. There are nuclear receptors for thyroid hormones, steroids, retinoids and vitamin D amongst others. Although these compounds are chemically different, they all share the property of being small, hydrophobic molecules that

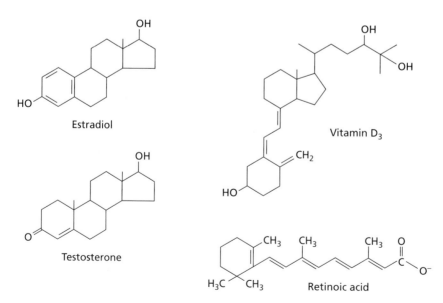

Fig. 8.8 Examples of ligands that can bind to nuclear receptors and influence their DNA-binding and gene regulation properties.

are able to diffuse across the cell membrane (Fig. 8.8). Once inside the cell, they are bound by the appropriate protein receptor, through which they exert their effects on the transcription of specific sets of genes.

All of the protein receptors bind to specific (different) DNA sequence elements as protein heterodimers and all use CBP/p300 as a coactivator. Two members of the nuclear receptor family, namely the retinoic acid receptor (RAR) and the thyroid receptor, provide striking examples of how the counterbalancing activities of HATs and HDACs can be used as part of an on/off gene switch. We will focus on the RAR as an example. This receptor has a particular relevance to human disease because its mutant forms have been implicated directly in the genesis and behaviour of human tumours.

When bound to its ligand, the RAR binds to defined DNA sequences as a dimer with a second, related protein designated RXR. *In vivo*, the ligand-bound dimer associates with a complex containing, amongst other components, the coactivators CBP/p300 and PCAF and activates transcription. Chromatin ImmunoPrecipitation (ChIP) experiments with antibodies to acetylated H3 and H4 have shown that up-regulation of transcription is accompanied by a transient increase in histone acetylation on nucleosomes close to the promoters of RAR-responsive genes. The mechanisms involved are just as illustrated earlier in Figs 8.4 and 8.5. The genes that are activated will depend on the cell type and its stage of differentiation, but in general terms they will be genes that are necessary for differentiation, or possibly programmed cell death (apoptosis). Conversely, failure to express RA-inducible genes encourages cell growth and proliferation.

It is interesting that, *in vitro*, the RAR/RXR heterodimer is able to bind to nucleosomal DNA almost as well as to naked DNA. In addition, retinoic acid has little effect on the binding of the dimer to chromatin *in vitro*. However, RA receptors can induce alterations in the organization of regularly spaced nucleosomes assembled over the RARβ2 promoter, often used as a model promoter in experiments to test RAR function. These results suggest that gene activation by the RAR/RXR heterodimer may involve reorganization of nucleosome arrays rather than displacement or disruption of individual, positioned nucleosomes. It may be that this reorganization of chromatin involves the chromatin remodelling complexes described earlier. There is strong evidence that activation by the glucocorticoid receptor (a closely related member of the nuclear receptor family), and possibly by RAR as well, requires the action of the mammalian SWI/SNF homologue, the BRG1 complex. Such findings make it likely that there is a hierarchy of interactions between different coactivator complexes and different chromatin modifying enzyme complexes. The enzymes recruited may vary from one cell type to another or even from one gene (or chromatin environment) to another.

The RAR/RXR heterodimer provides a good example of how the same receptor heterodimer can recruit different enzyme complexes. In the presence

(a) + ligand

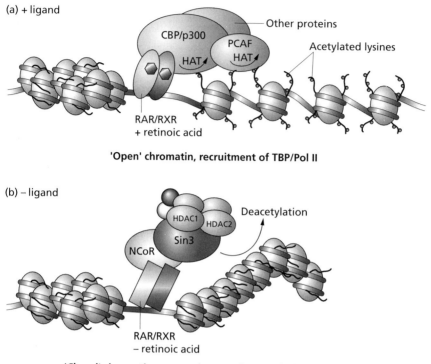

'Open' chromatin, recruitment of TBP/Pol II

(b) – ligand

'Closed' chromatin, no recruitment of transcription machinery

Fig. 8.9 Retinoic acid binding determines whether the RAR/RXR heterodimer recruits an activating histone acetyltransferase (HAT) complex or a repressing histone deacetylase (HDAC) complex. When the RAR/RXR heterodimer binds its ligand (retinoic acid) its shape changes. The liganded and unliganded receptors have the same DNA binding ability and specificity, but whereas the liganded receptor binds CBP/p300 HAT complex (a) the unliganded receptor binds the Sin3 deacetylase complex (b).

of ligand, RAR/RXR recruits the CBP/p300 HAT and associated proteins (Fig. 8.9a). In the absence of ligand, RAR/RXR heterodimers can still bind to their cognate DNA sequences both *in vitro* and in transiently transfected constructs, but they now actively *suppress* transcription by recruiting the Sin3 HDAC co-repressor complex (Fig. 8.9b). The evidence for this model comes largely from the use of *in vitro* techniques, and evidence that RAR binds specifically to DNA sites *in vivo* in the *absence* of ligand is still not available. Nonetheless, the model is consistent with the effect of HDAC inhibitors on transformed cells (see below) and is, at the very least, a good working model with which to try and make sense of a complex situation. It is worth emphasizing again that there is a strong biological justification for this type of model. Genes that play a key role in developmental or differentiation decisions, like those involved in cell cycle progression (they may sometimes be the same genes), must be tightly controlled. Their up- and down-regulation must occur within a specific and tightly defined time-frame. Simply removing the silencing complex and waiting patiently for the appropriate transcription factors to arrive, or removing the inducer and waiting for transcription to die down, may not be sufficiently accurate. A ligand-mediated, on/off switch of the type described seems much more appropriate.

Chromatin and cancer

The identification of enzymes that modify and remodel chromatin has taken our understanding of the role of chromatin in gene expression an important step forward. It has opened up new avenues for research, posed new questions and given us potentially important handles with which to manipulate gene expression, either for experimental or therapeutic reasons. But what has perhaps generated more interest and excitement than anything else in the field is the realization that these chromatin-modifying complexes can play a central role in the formation and progression of human cancer.

Human cancers often result from chromosome translocations. In many cases, these have been shown to alter either the structure or the pattern of expression of key growth-regulating genes. The most extensively studied example, and one that has become a paradigm for subsequent studies on chromosomes and cancer, is the 8:22 translocation found in the white blood cells of most patients with chronic myeloid leukaemia (CML). This translocation results in a small, partially deleted chromosome 22 (the so-called 'Philadelphia' chromosome) that contains a small piece from the end of chromosome 8. The translocated piece of chromosome 8 carries *ABL*, a gene that encodes a tyrosine kinase involved in signal transduction across the plasma (cell) membrane. On the Philadelphia chromosome, *ABL* is fused to a transcribed region designated the breakpoint cluster region (BCR). This results in a BCR/ABL fusion protein that is not responsive to the cellular controls that act on the normal ABL tyrosine kinase. The consequent unregulated growth of myeloid cells leads to CML. Note that the translocation that gave rise to the fusion gene occurred in a somatic cell, not in a meiotic (germ) cell. It is a form of somatic mutation. This is true of the majority of cancer-associated translocations, although there are exceptions where *meiotic* rearrangements cause inherited predisposition to specific cancers. There are now many examples of cancer-specific translocations, the majority of them associated with leukaemias or sarcomas. An increasing number of these are being found to disrupt genes encoding components of the coactivator and corepressor complexes containing HATs and HDACs.

The retinoic acid receptor and acute promyelocytic leukaemia

Two reciprocal translocations associated with certain types of acute promyelocytic leukaemia (APL) give rise to hybrid proteins involving the retinoic acid receptor, RAR (specifically the RAR *alpha* subunit). The two other proteins involved are PML (for ProMyelocytic Leukaemia protein) in cells with a t(15;17)

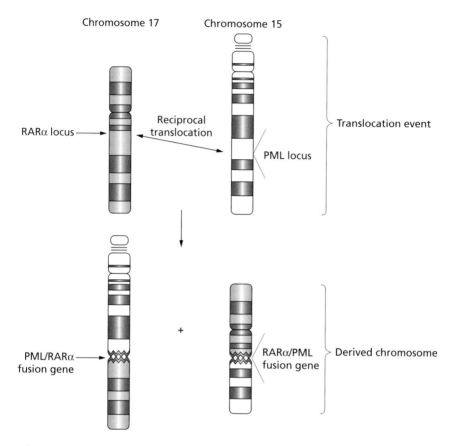

Fig. 8.10 A reciprocal translocation between human chromosomes 15 and 17 produces two novel fusion genes.

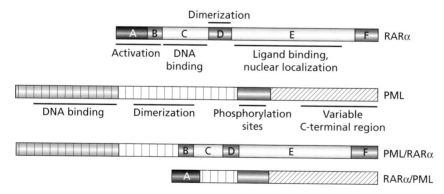

Fig. 8.11 Domain structures of the RARα and PML proteins and the novel fusion proteins found in ProMyelocytic Leukaemia (PML). (Adapted from Grignani *et al.*, 1994. *Blood* **83**: 10.)

reciprocal translocation, and PLZF (for promyelocytic leukaemia zinc finger protein) in cells with a t(11;17) reciprocal translocation. The PML/RARα fusion was the first to be identified and characterized. The translocation involved and the two derived chromosomes are shown in Fig. 8.10 and the structure of the resulting fusion proteins in Fig. 8.11. The function of the PML protein remains uncertain but its protein structure provides some clues. First, it contains a type of zinc finger motif (the RING finger) that is found in many proteins whose function involves DNA binding (e.g. the protein involved in DNA recombination, RAG-1). Second, it has a region of multiple α-helices with the potential to form a coiled-coil structure characteristic of dimerization interfaces. It resembles the structure involved in dimerization of the RAR itself. Finally, it has a domain rich in serine and proline residues that contains sequences likely to be recognized and phosphorylated by casein kinase II (CKII). This kinase modifies many transcription factors, often altering their functional properties. Of the two fusion proteins resulting from the reciprocal translocation, the larger, PML/RARa, contains most of the functionally important domains from the two proteins (Fig. 8.11) and is the one that has been shown to be influential in the pathogenesis of APL. The role of the smaller fusion product, if any, remains uncertain.

The story regarding the PLZF/RAR fusion is very similar. The PLZF locus is on chromosome 11 and the protein product is characterized by a motif containing nine zinc fingers and resembling a region in the transcription factor encoded by the *Drosophila* gap gene *Krüppel* (see Chapter 11). A t(11;15) translocation results in a fusion gene encoding a protein with the C-terminal part of PLZF, including two of the zinc fingers and an N-terminal region that contains most of RARα (Fig. 8.12). Exactly the same part of RARα is present in the PML/RARα and the PLZF/RARα fusion proteins.

The PML/RARα and PLZF/RARα proteins are both able to bind retinoic acid response elements (RARE) and to silence the genes controlled by these elements in the absence of ligand. However, this silencing continues even in the presence of physiological concentrations (10^{-9}–10^{-7} M) of retinoic acid. The hy-

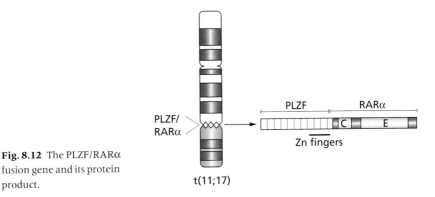

Fig. 8.12 The PLZF/RARα fusion gene and its protein product.

brid proteins are still able to bind ligand, but are unable to undergo the structural change that results in release of the corepressor complex and subsequent binding of the CBP/p300 coactivator complex (Fig. 8.9). The failure to activate genes required for normal progression down the pathway of myeloid differentiation results in a block in differentiation, inappropriate cell growth and, ultimately, leukaemia.

In general terms, the behaviour of the PML–RAR and PLZF–RAR hybrid proteins is very similar. However, studies in cultured cell lines revealed an important difference, namely that the corepressor complex is released from PML–RARα, but not from PLZF–RARα in the presence of elevated concentrations of retinoic acid. This may seem at first to be a rather insignificant experimental detail, but what makes it important is that the concentrations involved (10^{-7} to 2×10^{-5} M), although well above those found *in vivo*, are within the pharmacological (i.e. therapeutic) range. Thus, the experimental findings raised the possibility of a new approach to therapy for APL. The predictions have been borne out in practice, and patients with the PML/RARα variant of APL respond well to treatment with retinoic acid; the block in differentiation is released and the leukaemia goes into remission. Unfortunately, the remission usually lasts for only a year or so, after which the leukaemia returns in an RA-resistant form. In contrast, patients with the PLZF/RARα variant do not respond at all to RA therapy. The lack of RA responsiveness in patients with the PLZF–RAR variant results from the fact that PLZF itself can recruit the HDAC-containing corepressor complex. Even if the corepressor complex bound to the RAR component is released by RA, that bound to the PLZF component will not be.

Our new understanding of the mechanism of gene repression by the RAR and the ways it is disrupted in leukaemias gives hope that alternative therapies may be devised for those patients in which RA alone is ineffective, or who have become unresponsive to RA after prolonged treatment. One promising approach is to use HDAC inhibitors either alone or in combination with RA. Experiments have shown that the action of the PLZF-associated corepressor complex is sensitive to the HDAC inhibitor TSA and clinical trials are under way with another HDAC inhibitor, sodium phenylbutyrate.

The AML1 and ETO proteins

The gene most frequently disrupted in chromosome translocations associated with acute myeloid leukaemia is located on chromosome 21 and has been named AML1. It encodes a transcription factor necessary for activating genes that are essential for the differentiation of haematopoietic cells. These include the genes for growth factors such as interleukin-3 and granulocyte-macrophage colony stimulating factor (GMCSF) and for specialized enzymes such as myeloperoxidase. The AML1 protein has a highly conserved DNA binding domain homologous to that of the *Drosophila* transcription factor RUNT and its activation domain can recruit the HAT CBP/p300.

One of the most common chromosome translocations in AML is t(8;21) in which the N-terminal 177 amino acids of AML1 are fused with almost the whole of a gene named ETO (for Eight Twenty One, a handy reminder of the translocation that brought it to our attention). The hybrid protein encoded by the *AML1/ETO* gene retains the DNA-binding region of AML1, but has lost its ability to bind p300. To make matters worse, the ETO protein has the ability to bind specifically to the corepressor N-Cor and thereby recruit N-Cor/SMRT/HDAC corepressor complexes of the type described earlier. So, genes essential for correct differentiation of haematopoietic cells, far from being activated, are actively repressed (see Fig. 8.9). This results in the accumulation of cells blocked part-way down a differentiation pathway and, eventually, leukaemia. Unfortunately, the ability of the AML1–ETO hybrid protein to prevent differentiation is not effectively reversed by HDAC inhibitors. This may be because the catalytic activity of the deacetylase is not required for the function of this particular corepressor complex. Alternatively, it could be that the catalytic site in the hybrid protein is resistant to the deacetylase inhibitors tested thus far. Perhaps one of the more recently described inhibitors will be more effective. The answer to this question will be important in deciding upon the most promising approaches to therapy for this particular type of cancer.

The two examples described illustrate particularly well how disruption of the on/off switching mechanism that regulates genes essential for differentiation can lead to human leukaemias. These well-studied diseases and the specific translocations involved have proved to be particularly amenable to experimental analyses and the results have been quick in coming. However, it is likely that other human cancers, and other diseases, result from disruption of events mediated by corepressor/coactivator complexes. For example, yet another subtype of AML is characterized by a translocation, t(8;16)(p11;p13), that fuses the gene for CBP with MOZ, a gene encoding a protein with strong homology to known HATs (but whose *in vivo* function is not yet known).

Further Reading

ATP-powered remodelling complexes

Cairns, B. R., Lorch, Y., Li, Y. *et al.* (1996) RSC, an essential, abundant chromatin-remodeling complex. *Cell*, **87**: 1249–1260.

Côté, J., Quinn, J., Workman, J. L. & Peterson, C. L. (1994) Stimulation of GAL4 derivative binding to nucleosomal DNA by the yeast SWI/SNF complex. *Science*, **265**: 53–60.

Deuring, R., Fanti, L., Armstrong, J. A. *et al.* (2000) The ISWI chromatin-remodeling protein is required for gene expression and the maintenance of higher order chromatin structure in vivo. *Mol. Cell*, **5**: 355–365.

Krebs, J. E., Fry, C. J., Samuels, M. L. & Peterson, C. L. (2000) Global role for chromatin remodeling enzymes in mitotic gene expression. *Cell*, **102**: 587–598.

Peterson, C. L. & Workman, J. L. (2000) Promoter targeting and chromatin remodeling by the SWI/SNF complex. *Curr. Opin. Genet. Dev.*, **10**: 187–192.

Tyler, J. K. & Kadonaga, J. T. (1999) The dark side of chromatin remodelling: repressive effects on transcription. *Cell*, **99**: 443–446.

Remodelling mechanisms

Brehm, A., Längst, G., Kehle, J. *et al.* (2000) dMi-2 and ISWI chromatin remodelling factors have distinct nucleosome binding and modilization properties. *EMBO J.*, **19**: 4332–4341.

Côté, J., Peterson, C. L. & Workman, J. L. (1998) Perturbation of nucleosome core structure by the SWI/SNF complex persists after its detachment, enhancing subsequent transcription factor binding. *Proc. Natl. Acad. Sci., USA*, **95**: 4947–4952.

Hirschhorn, J. N., Brown, S. A., Clark-Adams, C. D. & Winston, F. (1992) Evidence that SNF2/SWI2 and SNF5 activate transcription in yeast by altering chromatin structure. *Genes Dev.*, **6**: 2288–2298.

Imbalzano, A. (1998) SWI/SNF complexes and facilitation of TATA binding protein:nucleosome interactions. *METHODS: A Companion to Methods in Enzymology*, **15**: 303–314.

Logie, C., Tse, C., Hansen, J. C. & Peterson, C. L. (1999) The core histone N-terminal domains are required for multiple rounds of catalytic chromatin remodeling by the SWI/SNF and RSC complexes. *Biochemistry*, **38**: 2514–2522.

Whitehouse, I., Flaus, A., Cairns, B. R. *et al.* (1999) Nucleosome mobilization catalysed by the yeast SWI/SNF complex. *Nature*, **400**: 784–787.

HATs and deacetylases

Brownell, J. E., Zhou, J., Ranalli, T. *et al.* (1996) Tetrahymena histone acetyltransferase A: a homologue to yeast Gcn5p linking histone acetylation to gene activation. *Cell*, **84**: 843–851.

Kochbin, S., Verdel, A., Lemercier, C. & Seigneurin-Berny, D. (2001) Functional significance of histone deacetylase diversity. *Curr. Opin. Genet. Devel.*, **11**: 162–166.

Marmorstein, R. (2001) Protein modules that manipulate histone tails for chromatin regulation. *Nature Reviews*, **2**: 422–432.

Ng, H. H. & Bird, A. (2000) Histone deacetylases: silencers for hire. *Trends Biochem. Sci.*, **25**: 121–126.

Taunton, J., Hassig, C. A. & Schreiber, S. L. (1996) A mammalian histone deacetylase related to the yeast transcriptional regulator Rpd3p. *Science*, **272**: 408–411.

Utley, R. T., Ikeda, K., Grant, P. A. *et al.* (1998) Transcriptional activators direct histone acetyltransferase complexes to nucleosomes. *Nature*, **394**: 498–502.

Effects on transcription (mutational analysis)

Holstege, F. C. P., Jennings, E. G., Wyrick, J. J. *et al.* (1998) Dissecting the regulatory circuitry of a eukaryotic genome. *Cell*, **95**: 717–728.

Chromatin and cancer

Cameron, E. E., Bachman, K. E., Myöhänen, S., Herman, J. G. & Baylin, S. B. (1999) Synergy of demethylation and histone deacetylase inhibition in the re-expression of genes silenced in cancer. *Nature (Genetics)*, **21**: 103–107.

Grignani, F., De Matteis, S., Nervi, C. *et al.* (1998) Fusion proteins of the retinoic acid receptor-α recruit histone deacetylase in promyelocytic leukaemia. *Nature*, **391**: 815–818.

He, L.-Z., Guidez, F., Tribioli, C. *et al.* (1998) Distinct interactions of PML–RARα and PLZF–RARα with corepressors determine differential responses to RA in APL. *Nature (Genetics)*, **18**: 126–.

Jones, P. A. & Laird, P. W. (1999) Cancer epigenetics comes of age. *Nature (Genetics)*, **21:** 163–167.

Lin, R. J., Nagy, L., Inoue, S. *et al.* (1998) Role of the histone deacetylase complex in acute promyelocytic leukaemia. *Nature*, **391:** 811–814.

Chapter 9: Heterochromatin

Introduction

In a series of microscopical studies of plant and animal cells published between 1928 and 1934, the German cytologist Heitz described how defined regions of specific chromosomes could be visualized as discrete, strongly staining ('heteropycnotic') patches in interphase nuclei. The intensely staining patches were termed *heterochromatin*, to distinguish them from the relatively pale staining chromatin distributed through the rest of the nucleus, which was termed *euchromatin*. The fact that these patches stained as intensely in interphase as they did when formed into condensed, prophase chromosomes suggested that they represented chromatin that remained relatively condensed during interphase. Genetic studies in *Drosophila* rapidly revealed two additional properties of the regions of heterochromatin, as originally defined. First, they contained relatively few genes, and, second, very little crossing over occurred in these regions during meiosis. A little later, with the advent of radiolabelling and autoradiographic techniques, it was shown that heterochromatin replicated later in S-phase than euchromatin. These properties have stood the test of time, in that they have been found to be true of regions identified as heterochromatic in a large number of different tissues and cell types. However, 60 years of experimentation has revealed (amongst many other things) that none of them is true of all types of heterochromatin in all cell types and all developmental situations.

The two properties that are most frequently associated with heterochromatin are genetic inactivity and a tendency to remain relatively condensed throughout the cell cycle. These are not characteristics that inspire immediate excitement or interest and, for many years, heterochromatin was something of a backwater in chromatin research. This has now changed, largely because of the realization that blocks of heterochromatin can influence the transcription of genes elsewhere in the genome and that adoption of a 'heterochromatin-like' chromatin structure can switch off specific genes, groups of genes or even whole chromosomes. These constitute vitally important mechanisms of gene regulation.

Unfortunately, the increased interest in heterochromatin led to the term becoming so widely applied to different types of chromatin that any simple definition became, at best, incomplete and at worst downright misleading. Even the two properties listed at the outset are not true of all types of chromatin to which the designation 'heterochromatin' has been applied, nor are the properties themselves as straightforward as might at first appear. The first property, genetic inactivity, can mean either that heterochromatin contains very few genes (although those that *are* present may be transcriptionally active) or that genes are present at normal frequency but are rendered transcriptionally inactive. As outlined below, both are true of different types of heterochromatin. The second property, a condensed (or compacted) structure, also requires clarification. The idea that heterochromatin is condensed derives primarily from the observation that it stains relatively strongly with commonly used DNA stains in interphase cells. Because heterochromatin stains as strongly as metaphase chromosomes, which are undoubtedly condensed, it has been assumed that its strong staining is also a reflection of chromatin condensation. However, in cases where this has been tested, either by measuring susceptibility to digestion with nucleases or, more recently, by three-dimensional imaging, the results have not been consistently in favour of increased condensation. What they *have* consistently suggested is that the structure of heterochromatin is somehow *different* from that of bulk chromatin, although as yet we have no molecular description of this difference. It may well vary between species, cell types or even different stages of the cell cycle. It seems likely that, as with other adjectives applied to heterochromatin in general, 'condensed' is just too simple a term.

α and β heterochromatin in *Drosophila*

The difficulty of assigning definitive properties to heterochromatin became apparent even in the very early experiments. As a result of his work with polytene chromosomes from the salivary gland cells of *Drosophila* larvae, Heitz concluded that these cells contain two types of heterochromatin, which he termed α and β. They differed in their staining properties, their location within the nucleus and in the extent to which they were replicated during chromosome polytenization (i.e. DNA amplification). Both types of heterochromatin were located in the chromocentre, a region where the centromeres of the four polytene chromosomes associate, but could be distinguished by their staining properties. The α-heterochromatin was present in a small, intensely stained region, while the β-heterochromatin was more diffusely spread along the proximal regions of the chromosome arms (Fig. 9.1, see also the immunostained polytene chromosome spreads in Fig. 5.17). It soon became clear that polytene chromosome spreads give a rather misleading impression of the amount of heterochromatin in the *Drosophila* genome. Staining of *Drosophila* metaphase chromosomes shows that they contain blocks of heterochromatin far larger (proportionally) than those seen in interphase polytene chromosomes (Fig. 9.2). The Y chromosome is almost entirely heterochromatic and does not feature at all in polytene chromosome spreads. The most likely explanation for this discrepancy is that at least some of the heterochromatic regions do not endoreduplicate (i.e. form multistranded, polytene chromosomes) as efficiently as the rest of the genome. On the whole, they remain diploid and stay hidden in the patch of α-heterochromatin. However, it seems that some parts of the Y chromosome can polytenize very effectively, although they remain hidden in the chromocentre.

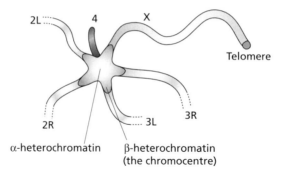

α-heterochromatin   β-heterochromatin
(the chromocentre)

Fig. 9.1 The chromocentre in a *Drosophila* polytene nucleus. The centromeres of the large chromosomes 2, 3 and X cluster together to form the chromocentre. The centromeres of chromosomes 2 and 3 are near the middle of each chromosome whereas that of X is at one end. The chromosome arms either side of the centromere are designated R and L. Chromosome 4 has extensive regions of heterochromatin and remains closely associated with the chromocentre. *Drosophila* chromosomes have extensive regions of repetitive DNA that do not undergo the several rounds of endoreduplication that leads to polytene chromosome formation. These regions, designated α heterochromatin, are under-represented in polytene chromosomes. The diagram should be compared with the stained polytene chromosome spread shown in Fig. 5.17.

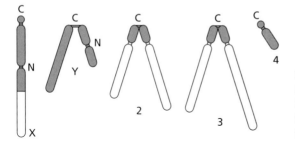

Fig. 9.2 Heterochromatin and euchromatin in *Drosophila* metaphase chromosomes. Regions of heterochromatin and euchromatin are interspersed across chromosome 4. (Redrawn from Ashburner 1989.)

The internal structure of this cytologically rather nondescript region is likely to be complex.

β-heterochromatin provides the first example of the danger of attempting to generalize about the properties of heterochromatin. Despite its cytological appearance, β-heterochromatin is transcriptionally active. It contains at least three multicopy gene families, including the highly active rRNA genes, whose transcription leads to formation of a nucleolus, together with several single copy genes, mutation of which leads to a variety of effects on viability, fertility and morphology. As discussed in more detail later, β-heterochromatin and the genes it contains have their own characteristic properties, including distinctive chromatin structural features and the frequent occurrence of repetitive DNA sequence elements that are rare or absent in euchromatin.

Whatever its structure might be, chromocentric β-heterochromatin is certainly the most prominent type of heterochromatin in polytene cells. Additional, smaller but functionally important, regions of heterochromatin are found at telomeres and throughout chromosome 4. This small chromosome is unusual in having patches of heterochromatin throughout its length, interspersed with active, normally regulated genes. It is usually located close to the chromocentre.

Facultative and constitutive heterochromatin

Heterochromatin can be classified into one or the other of two major subtypes, namely *facultative* and *constitutive*. The former can be defined as chromatin that has all the characteristics of euchromatin in terms of gene density and DNA sequence characteristics but which, under specific *in vivo* circumstances, can adopt the structural and functional properties of heterochromatin. These properties include intense staining, giving a 'condensed' appearance, replication late in S-phase and loss of transcriptional activity. The best studied example of facultative heterochromatin is the inactive X chromosome (Xi) in the somatic cells of female mammals. This process provides a fascinating and important example of the mechanisms by which chromatin can exert a long-term influence on genetic activity and is described in more detail in Chapter 12.

Constitutive heterochromatin can be defined as that which has DNA sequence motifs, particularly repetitive elements, that distinguish it from euchromatin (see below). However, it is important to emphasize that these sequence motifs do not guarantee that a particular stretch of DNA is *always* heterochromatic. The DNA sequence motifs provide only the *potential* to form heterochromatin. Whether or not this potential is realized depends on the availability of other components in the nuclear milieu. An important insight that has come out of heterochromatin research over the last few years has been the realization that heterochromatin is a dynamic and flexible entity that can change in parallel with changes in growth characteristics or differentiation.

Heterochromatin DNA

If DNA is isolated from eukaryotic cells, freed from proteins and fractionated by centrifugation through a caesium chloride density gradient, it resolves into a series of peaks. An example is shown in Fig. 9.3. This resolution occurs because centrifugation under these conditions separates DNA molecules not by size but by their density. During centrifugation, each DNA molecule sediments to that point in the gradient at which its density equals that of the caesium chloride solution, and there it stays. Density is dependent, primarily, on the base composition of the DNA, with DNA molecules rich in AT base pairs having a lower density than those rich in GC base pairs. The single major peak (known as 'main band' DNA) comprises most of the genomic DNA and contains a mixture of sequences made up of coding and noncoding DNA from different parts of the genome. Its density will reflect the average AT:GC ratio of genomic DNA, which

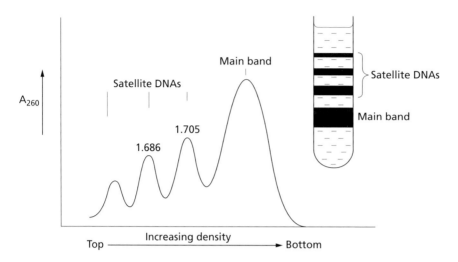

Fig. 9.3 Separation of main band and satellite DNAs from *D. melanogaster* by centrifugation on a caesium chloride gradient. On prolonged high-speed centrifugation, caesium chloride ($CsCl_2$) forms a concentration gradient in the centrifuge tube, lowest density at the top and highest at the bottom. DNAs migrate to that position in the tube that corresponds to their density. The density of a piece of DNA is largely independent of its size, but very sensitive to base composition. Fractionation of the gradient allows the position of the various bands to be determined (by optical density measurement) and the different DNAs to be isolated for further analysis.

can vary from one species to another. In *D. melanogaster*, the AT content of main band DNA is 57% and the median density (usually designated by ρ) is $1.701\,g\,cm^{-3}$. In contrast, the minor ('satellite') peaks contain DNA molecules whose base composition differs radically from that of main band DNA. The reason for this difference is that the satellite DNAs are made up largely of simple sequences repeated, in tandem, many times over. For example, the DNA in the peak sedimenting at $\rho = 1.672$ consists predominantly of repeats of the five base sequence AATAT. Such repeats are written as $(AATAT)_n$, where n can be up to several thousand. The satellite DNAs account for about 21% of the haploid genome of *D. melanogaster*. It is important to note that the composition of satellite DNAs is remarkably species specific. The satellite DNAs of *D. melanogaster* are quite different from those of the sibling species *D. simulans*.

The advent of DNA cloning and sequencing has made it possible to study the DNAs within each peak in more detail. Each peak consists of a mixture of sequences, the most frequent of which are listed in Table 9.1. *In situ* hybridization with labelled satellite DNA can be used to show how the different satellites are distributed along chromosomes. In all cases, the satellite DNAs are localized, predominantly, in those regions defined, by cytological and genetic approaches, as constitutively heterochromatic. Table 9.1 shows that each satellite DNA has its own, characteristic distribution among the five different *Drosophila* chromosomes. Two closely related satellite sequences (AAGAG and AAGAGAG) seem

Table 9.1 Satellite DNAs in *Drosophila melanogaster*.

Satellite		Satellite DNA (Mb) per chromosome				
Sequence	Density	X	Y	2	3	4
AATAAAC	1.669	0	1.6	0	0	0
AATAT	1.672	0.60	5.8	0.01	0.63	2.7
AATAC	1.680	0	3.5	0	0	0
AATAACATAG	1.686	0	0	1.9	1.6	0
AATAGAC	1.688	0	1.6	0	0	0
359 bp	1.688	11	0	0	0	0
AAGAC	1.689	0.08	8.5	1.8	0.11	0
	1.701					
AAGAG	1.705	1.2	7.2	5.5	1.1	0.17
AAGAGAG	1.705	0.27	1.8	1.7	0.14	0.10
Total satellite		16.1	32.5	11.1	3.6	3.0
Total DNA		40	34	62	66	6

to confer particular staining characteristics on the chromosome regions in which they are located, giving rise to the staining pattern known as N-banding. The *Drosophila* Y chromosome, found only in males, is mostly made up of satellite DNA and is heterochromatic, containing only a small number of genes necessary for male fertility. Chromosome 4, by far the smallest *Drosophila* chromosome, is 50% satellite DNA (mostly AATAT repeats) but is also rich in middle-repetitive sequences and contains significant numbers of genes, some of which are essential for viability.

In addition to the satellite DNAs, *Drosophila* heterochromatin is also rich in transposable elements (TEs). These can be categorized into 30–50 separate families and account for about 10% of genomic DNA in *Drosophila*. Many of these sequences are derived from retroviruses and many contain long-terminal repeats characteristic of such viruses and regions homologous to the viral *pol* (reverse transcriptase) and *gag* (polyprotein) genes. Many TEs move around the genome by being transcribed, making a DNA copy of the resulting RNA with reverse transcriptase, and then reinserting the DNA copy at a new location. A careful analysis of the distribution of TEs across *Drosophila* metaphase chromosomes was carried out by *in situ* hybridization with fluorescently labelled DNA probes. The results showed that the elements were rare in euchromatin, but were scattered across centric heterochromatin, with each element (of the 11 tested) showing its own characteristic distribution. Examples are shown in Fig. 9.4. These distributions were found to be highly conserved between different *Drosophila* species.

Two important points emerge from all this. The first is that centric heterochromatin is a mosaic of DNA elements; satellite sequences, transposable ele-

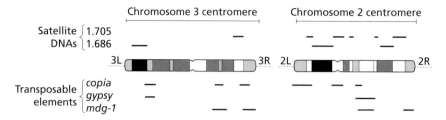

Fig. 9.4 Distribution of satellite DNAs and transposable elements across the centric heterochromatin of *Drosophila melanogaster* chromosomes 2 and 3. (Prepared from data presented in Pimpinelli *et al.*, 1995.)

Table 9.2 Classes of satellite (tandemly repeated) DNA in human cells.

Class	Size of blocks	Family	Repeat unit size (bp)	Major locations
(Major) Satellite	0.1 Mb to >2 Mb	satellite 1	25–48	Centromeric and other heterochromatin; AT rich
		satellites 2, 3	5	Most chromosomes
		alpha satellite	171	Centromeric heterochromatin, all chromosomes
		beta satellite	68	Centromeric heterochromatin, chrom's 1,9,13,14,15,21,22,Y
Minisatellite	0.1–20 kb		6–24	At or near telomeres
Microsatellite	often <150 bp		1–4	All chromosomes

ments and the occasional gene (see below). The second is that different blocks of heterochromatin have their own, characteristic composition in terms of these elements. These differences may be functionally important.

Satellite DNAs are a common characteristic of heterochromatin in many species, including our own. As in flies, human satellite DNAs come in varying sizes and frequencies and have their own, characteristic chromosomal locations, although the analysis is not nearly so detailed as in *Drosophila*. Human satellite DNAs are described in Table 9.2 and an example of their distribution around the centromere of chromosomes 9 and 21 is shown in Fig. 9.5. Transposable elements are also found in the human genome and fall into two major groups, SINES (short interspersed nuclear elements) and LINES (long interspersed nuclear elements). The most frequently occurring members of each group are shown in Fig. 9.6. Both sequences are very variable and LINE-1 sequences can vary in length over a five-fold range; the *average* size of LINE-1 sequences is only 1.4 kb, compared with 6.1 kb for the full-length element. Both sequences are distributed throughout the genome, without the preference for heterochromatin shown by *Drosophila* transposable elements. However, the distribution of these elements is not random. The AT-rich LINES are relatively enriched in the G-bands of metaphase chromosomes and the GC-rich SINES in R-bands (see Chapter 5). Both are frequently transcribed.

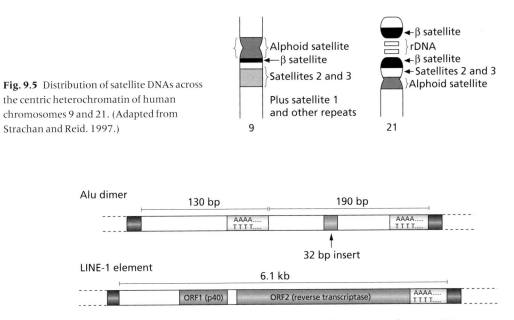

Fig. 9.5 Distribution of satellite DNAs across the centric heterochromatin of human chromosomes 9 and 21. (Adapted from Strachan and Reid. 1997.)

Fig. 9.6 Examples of full-length human SINE (Alu) and LINE-1 DNA elements. ORF=open reading frame.

Heterochromatin genes

About 40 genes in *Drosophila*, some of them essential for viability, have been mapped to regions of heterochromatin. These include the 18S and 28S ribosomal RNA genes (tandemly arrayed, multicopy genes, counted as one gene here) and 30 or so genes whose mutation results in defects in fertility, viability or morphology. These genes are organized rather differently to euchromatin genes in that their flanking regions and larger introns contain high densities of middle-repetitive DNAs similar to those located in other regions of β-heterochromatin. The gene *light*, an essential gene located in the heterochromatin of chromosome 2 and expressed in a number of cell types, is typical. The primary transcripts of some genes encoding fertility factors and located on the heterochromatic Y chromosome have primary transcripts larger than 1 Mb consisting predominantly of satellite or middle repetitive sequences. Heterochromatin genes not only tolerate the particular environment provided by heterochromatin, they *need* it for their expression. If they are translocated into euchromatin, their expression is reduced or lost, an interesting reversal of the more usual type of position effect discussed below. This behaviour must reflect idiosyncrasies in the control of these genes, possibly due to the presence of middle-repetitive DNA sequences (characteristic of β-heterochromatin) within and around them. The point that should be emphasized is that expression of these genes shows that RNA polymerases and basal transcription factors are not excluded from the regions of the nucleus containing heterochromatin.

Heterochromatin proteins

Despite the characteristic DNA sequences associated with heterochromatin in many different organisms, it seems that DNA content alone is not sufficient to give heterochromatin its distinctive properties. This is self evident in the case of facultative heterochromatin, where homologous genomic regions (with functionally identical DNA compositions) can be heterochromatic and euchromatic in the same cell. In the case of constitutive heterochromatin the evidence is largely circumstantial, deriving from observations showing that the appearance and behaviour of heterochromatin can vary depending on growth properties or stage of development. Most models of heterochromatin structure rely not on DNA sequence alone, but involve the interaction of specific proteins with DNA.

Histones in heterochromatin

Early experiments showed that the most common chromatin proteins, the histones, were present in the same proportions in heterochromatin and euchromatin and that, in each type, the nucleosome was the fundamental unit of DNA packaging. However, there are now known to be more subtle differences in the histone content of heterochromatin and euchromatin. Histones in heterochromatin and euchromatin differ in the way in which they are post-translationally modified by acetylation. By immunofluorescence microscopy with antibodies specific for the acetylated isoforms of histone H4, it has been shown that H4 is underacetylated in the heterochromatin of both mammals and *Drosophila*. Underacetylation of H4 (and usually the other core histones too) has become one of the most consistent properties of both facultative and constitutive heterochromatin across a wide range of species. Blocks of heterochromatin adjacent to the centromeres of human metaphase chromosomes, the facultative heterochromatin of the inactive X chromosome in female cells and the blocks of heterochromatin in both diploid and polytene *Drosophila* cells are all underacetylated (examples are shown in Plate 5.1 and Fig. 5.17).

An additional, more subtle, difference was revealed by labelling polytene chromosome squashes from *D. melanogaster* larvae with antibodies specific for H4 acetylated at particular lysine residues (i.e. lysines 5, 8, 12 or 16). While H4 in the β-heterochromatin of the chromocentre and chromosome 4 was underacetylated at lysine residues 5, 8 and 16, it was acetylated at lysine 12 at a level equal to, or above, that seen in euchromatic regions. This lysine-specific difference in acetylation argues against the possibility that H4 underacetylation is simply a *consequence* of heterochromatin formation. For example, it seems unlikely that heterochromatic regions would be inaccessible to the enzymes acetylating H4 at some lysines but not at others. The observation raises the interesting possibility that acetylation of H4 at lysine 12, in a region that is generally underacetylated, may have a specific role in heterochromatin assembly.

A second, residue-specific histone modification that may have a role in chromatin condensation, if not specifically in heterochromatin formation, is phosphorylation of histone H3 at serine 10. This occurs in late G2 as cells prepare for mitosis. Phosphorylation first appears in centromeric heterochromatin and then spreads throughout the chromosomes as the cell moves into mitosis. This post-translational change is just one of several, targeted post-translational histone modifications associated with chromatin condensation. The cell-cycle dependent combination of histone-specific and residue-specific acetylation, phosphorylation and methylation (see below) found on centromeric heterochromatin provides a good example of how combinations of histone modifications can provide a unique epigenetic mark associated with defined changes in chromatin structure and function.

Centromeric heterochromatin also contains a variant form of histone H3 in which the central, globular portion of the molecule is very similar to normal H3, but in which the N-terminal tail is quite different, sometimes extended. Such variants are found in humans and mice (CenpA), *Saccharomyces cerevisiae* (Cse4p), *Caenorhabditis elegans* (various) and *Drosophila* (Cid). These variants localize to centric heterochromatin even when expressed (by transient transfection) in cells of a different organism. Their functional significance is unclear, but is likely to be related to centromere function, rather than to formation of heterochromatin *per se*.

Non-histone proteins in heterochromatin

The gene regulatory protein GAGA factor has been located to centric heterochromatin by immunofluorescence microscopy. This is not unexpected as one of the major *D. melanogaster* satellite DNAs ($1.705 \, g \, cm^{-2}$) contains the GAGA motif (Table 9.1). GAGA factor may be important for heterochromatin formation and function; embryos in which it is improperly located during early development show severe mitotic defects. Very recently, the protein product of a gene designated *prod* (proliferation disrupter) has been shown to bind to the $1.686 \, g \, cm^{-2}$ satellite of *D. melanogaster* (Table 9.1). *Prod* binds the satellite repeats in a cooperative manner, and it has been suggested that the formation of protein multimers is an integral part of condensation of satellite DNA arrays.

The first heterochromatin binding protein to be identified and still the best characterized is heterochromatin protein 1 (HP1) from *Drosophila*. Immunolabelling of polytene chromosome squashes with antibodies to HP1 showed that the protein is located predominantly to regions of β-heterochromatin adjacent to the chromocentre and to chromosome 4 (Fig. 9.1). Its distribution in these regions parallels that of a DNA element called 'DrD', a moderately repeated sequence characteristic of β-heterochromatin. HP1 is also seen at sites along the chromosome arms and at some telomeres. It is significant that HP1 also localizes to small regions of heterochromatin translocated into the chromosome arms, showing that its binding is dependent on DNA sequence information rather than chromosomal location.

Chromodomain · Hinge · Chromoshadow domain

Fig. 9.7 The domain structure of heterochromatin protein HP1.

The amino acid sequence of HP1 contains motifs that provide clues to its function and possible mode of operation (Fig. 9.7). Of particular interest is a region designated the chromodomain (for *chromatin organization modifier*). This domain is also found in members of the polycomb group (PcG) gene family, whose protein products have essential roles in selective gene silencing during development (Chapter 10), and in a variety of other proteins involved, in one way or another, in modifying chromatin structure and regulating gene expression. The domain has been highly conserved through evolution and has been found in proteins from many species, including humans. HP1 and PcG proteins and their interactions are considered in more detail in Chapter 10, but it is important now to point out something that both HP1 and PcG proteins (with one exception) do *not* have, namely any of the various sequence motifs characteristic of DNA-binding proteins. Nor do they show any evidence of significant DNA binding activity in *in vitro* assays. In view of this, the location of HP1 to specific sites on chromosomes must be the result of interaction (direct or indirect) with proteins that *do* have DNA-binding ability.

Surprisingly, recent results have shown that one such protein may be a specifically modified form of histone H3. *In vitro* binding assays show that HP1 binds strongly to synthetic peptide corresponding to the N-terminal region of histone H3 (at least residues 1–16) *only* if it is methylated at lysine 9. HP1 does not bind to the unmethylated H3 peptide, or to an H3 peptide methylated at lysine 4. Two pieces of evidence suggest that this interaction may be significant *in vivo*. Firstly, an enzyme that methylates H3 specifically at lysine 9 (designated Suvar39h, for reasons that will become clear by the end of the chapter) locates to centric heterochromatin *in vivo*. In mice deficient in this enzyme, HP1 does not locate to heterochromatin as usual. Secondly, HP1 can be displaced from heterochromatin by treatment with high concentrations of methylated H3 peptide. HP1 binding to H3 methylated at lysine 9 (H3me9) can be abolished by mutation of its chromodomain. These are exciting results, but do not provide a complete answer. For example, there is a lot of H3me9 that is not associated with heterochromatin, and mice lacking Suvar39h have near normal levels of H3me9 in bulk histones. H3me9 *alone* cannot provide a heterochromatin-specific signal. We return to this point at the end of the chapter.

Proteins that interact with HP1

Using a variety of experimental approaches, several proteins have been identified that interact with *Drosophila* HP1, or its homologues in mice and humans. In some cases the interaction is direct, while in others the protein is part of an HP1-

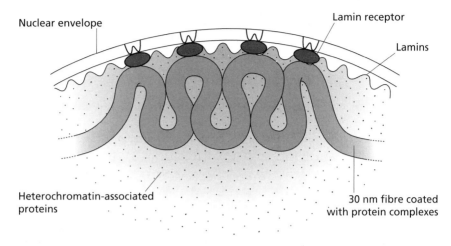

Fig. 9.8 Heterochromatin domains can be associated with the nuclear envelope and may occupy regions rich in heterochromatin-associated proteins such as HP1.

containing multiprotein complex. It is significant that HP1 has been shown to colocalize or to coimmunoprecipitate with components of the origin of replication complex (ORC) in *Drosophila*. This complex is responsible for the initiation of DNA replication at defined genomic sites. Replication late in S-phase is an almost universal property of heterochromatin and may even be necessary for heterochromatin formation. The HP1–ORC interaction may be just for targeting the replication machinery to heterochromatin, or may play a more subtle role in determining replication timing. Another significant interaction is with the lamin B receptor. This protein is embedded in the internal nuclear envelope and helps locate the network of lamins A, B and C at the nuclear periphery (Fig. 5.19). Its interaction with HP1 could explain the frequent localization of heterochromatin to this part of the nucleus.

Evidence confirming the existence of a family of nonhistone proteins necessary for heterochromatin assembly has come primarily from genetic experiments and no protein involved in heterochromatin formation has yet been described in such detail as HP1. However, with the current information we have, it is possible to construct a simple working model of heterochromatin formation, based on the assembly of multiprotein complexes. This is presented diagrammatically in Fig. 9.8.

Position effect variegation

Heterochromatin is not a suitable environment for the expression of most genes. Experiments in which genes are randomly inserted into the genomes of mice and flies have shown that when these genes are located in regions of heterochromatin (a frequent occurrence) they are not expressed. This occurs even when the transgene carries with it strong promoter and enhancer elements.

Only specific DNA elements termed boundary elements or locus control regions can enable a euchromatic gene to be expressed when surrounded by heterochromatin. (These DNA elements were described in Chapter 6.) So, blocks of constitutive heterochromatin have a major, often overriding, influence on the expression of euchromatin genes.

A more puzzling observation dates from early genetic experiments with *Drosophila*. In a small number of mutant flies, the mutant phenotype was expressed in some cells but not others. It was described as 'variegated'. These variegating mutants were shown to be caused by chromosome rearrangements that resulted in the movement of a euchromatin gene from its normal position to a site adjacent to a block of heterochromatin. The heterochromatin could be at the centromere, telomere or on chromosome 4. The mutant effect reflected not a change in the gene itself, but in its chromosomal location. The phenomenon was termed position effect variegation (PEV) and has fascinated and puzzled geneticists for over 60 years.

The best-studied example of PEV is that resulting from variegated expression of the *white* gene. This gene is required for the red eye colour normally seen in *Drosophila* and loss of function mutants result in flies with white eyes (hence the name of the gene). It is located near the distal end of the left arm of the X chromosome. Chromosome rearrangements that place the *white* gene adjacent to the heterochromatin surrounding the X centromere result in silencing of the gene in some cells but not others. In males who have only a single, rearranged X, this leads to a phenotype in which the eyes are a mosaic of red (wild type) or white (mutant) patches (Fig. 9.9). (The same phenotype is seen in XX females who have the rearranged X along with a second X carrying a nonfunctional *white* allele.) The extent of variegation can be measured by the relative abun-

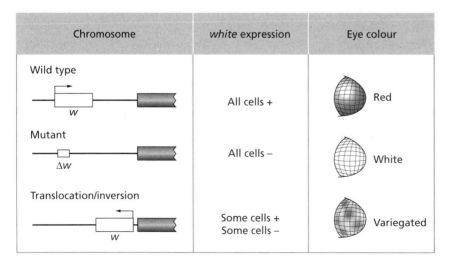

Fig. 9.9 Chromosome rearrangements that move the *white* (*w*) gene to a position close to a block of heterochromatin (dark shading) cause variegated expression.

dance of red or white areas. The ease with which this variegating phenotype can be studied is an important factor in the continuing popularity of the *white* gene as an experimental model. The pattern of variegation in the eye can best be interpreted in the following way. First, silencing of the *white* gene, where it occurs, happens relatively early in development and, second, once the gene has been silenced in a particular cell, then it *remains so in all the progeny of that cell*. In other words, the silent state is *stably inherited* from one cell generation to the next. It is this stable inheritance that gives rise to the patches of red and white, with each patch being derived from an original single cell in which the *white* gene was either on (active) or off (silent). Note that this form of inheritance is not based on DNA sequence, which is the same in the silent and active forms of the gene, but on some other method of information transfer. It falls into the general category of epigenetic inheritance.

Chromatin structure and PEV

The genetic evidence shows that genes translocated adjacent to heterochromatin can be silenced. In searching to explain this observation, two general models provide useful starting points. First, genes could be silenced because their chromatin structure changes, perhaps becoming more 'heterochromatin-like'. This requires some form of 'spreading' of the heterochromatin-like structure until it invades the adjacent gene. Alternatively, regions adjacent to heterochromatin could be sheltered from factors necessary for the transcription of euchromatic genes. This is an effect of intranuclear location and does not require change in their chromatin structure. As usual, the models are not mutually exclusive and both mechanisms may have a role to play (Fig. 9.10).

Early experiments suggested that changes in chromatin structure were involved. Deletion mutants with less than the usual number (100–150) of histone genes were remarkably healthy, but showed reduced PEV (i.e. less efficient silencing), suggesting that silencing depends on efficient chromatin assembly. In addition, if larvae were fed with the inhibitor of histone deacetylases, sodium butyrate, PEV was also suppressed, showing that treatments that disrupt the usual underacetylation of heterochromatin also disrupt PEV. (Note, however, that butyrate is unpleasant and toxic, even for flies.) Also, direct microscopical analysis of the banding structure in polytene chromosomes of euchromatic regions translocated near to heterochromatin showed increased condensation in some nuclei. More recent experiments have compared the chromatin structure of the heat shock gene *hsp26* inserted as transgenes into heterochromatin or euchromatin. Despite its strong promoter, the inducibility of *hsp26* is impaired in heterochromatin. It was found that the transgenes in heterochromatin were less sensitive to nuclease digestion and were packaged in a more regular array of nucleosomes than their euchromatic counterparts. So, the weight of evidence is that heterochromatin suppresses the transcription of adjacent euchromatic genes by altering their chromatin structure, although the results are by no

(a) Spreading model   (b) Nuclear relocation model

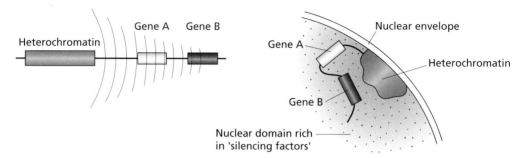

Fig. 9.10 Mechanisms by which heterochromatin might inactivate nearby genes. (a) In this model, a repressive chromatin structure, originating in a block of heterochromatin, spreads along the chromatin fibre. The extent of spreading varies from one nucleus to another, leading to variegated expression of nearby genes. Gene A will be inactivated more frequently than gene B. The model predicts that in nuclei in which gene B has been inactivated, gene A will always be inactive. (b) This model proposes that heterochromatin occupies nuclear domains rich in heterochromatin proteins and silencing factors. Genes located close to heterochromatin will inevitably be drawn close to these domains and may be inactivated. The model allows for the possibility that, in the same nucleus, a gene that is closer (in a genetic sense) to the heterochromatin can lie outside the silencing domain, whereas a more distal gene may lie within it. This will depend on the path of the chromatin fibre.

means unanimous. Experiments in which the accessibility of specific restriction enzyme sites was measured by ligation-mediated PCR failed to detect any change as a result of silencing by PEV. Perhaps the changes in chromatin structure associated with gene silencing are more subtle than simple models of heterochromatin would suggest.

Early observations on the mechanism of PEV

An early observation on factors that influence PEV is consistent with the simple, working models of heterochromatin presented in Figs 9.8 and 9.10 and introduces a new twist. This is that the extent of variegation in flies carrying a re-arranged *white* gene can be influenced by the amount of heterochromatin on other chromosomes. The most straightforward example of this is that in male flies carrying a Y chromosome, which is almost entirely heterochromatic, variegation is much *less* than in otherwise identical flies lacking a Y (i.e. the *white* gene is less frequently silenced). Furthermore, in flies bred to have two Y chromosomes, variegation is reduced even more. These observations can be explained by suggesting that the Y chromosome competes with other chromosome regions for some or all of the proteins necessary for heterochromatin formation. Because of this competition, heterochromatin is less likely to encompass the *white* gene and therefore the gene will be inactivated less frequently. These observations suggest that one or more of the heterochromatin-forming proteins are present in limiting amounts. If they were all present in excess, then the addition of more heterochromatin would be expected to have no effect.

Su(var)s and E(var)s; proteins that influence PEV

If some or all of the heterochromatin-forming proteins are present in limiting amounts, then reducing the amount of such a limiting protein by half would have a measurable effect on PEV. Such a reduction would be brought about by mutations in the genes encoding these proteins. In order to identify such genes, extensive surveys have been made for mutant flies in which PEV is either enhanced or suppressed, i.e. in which the balance between cells expressing the marker gene (usually *white*) and those not expressing it is shifted one way or the other. The rationale for these searches is that mutations that inactivate one or the other of the heterochromatin-forming proteins, even in single dose, will reduce (i.e. *suppress*) variegation while those that either increase their amount or improve their function will increase (i.e. *enhance*) variegation.

Surveys designed to identify mutations that modify PEV have yielded a rich harvest and over 100 such mutants have so far been identified. As would be expected, suppressor mutants, termed Su(var)s, are more frequent than enhancers, termed E(var)s. Note that a typical Su(var) can act as an *enhancer* of PEV in flies that have been bred to have three copies of the gene instead of the usual two. It will act as a 'triplo-enhancer' and 'haplo-suppressor'. This is entirely consistent with our working model, in that three copies of the gene will give a 50% increase in protein and a corresponding increase in the efficiency of heterochromatin formation. One small complication is that although Su(var)s and E(var)s can act as suppressors of variegation for genes located within centric heterochromatin or on chromosome 4, they do not always affect the expression of genes located adjacent to telomeres. Telomeric heterochromatin must be different, in some important functional sense.

Of course, many of the mutants identified in these screens will be a result of changes in genes only indirectly involved in PEV and heterochromatin formation. This is inevitable in screening for effects on such a complex phenotype as PEV. Genes have been cloned from some of the most promising mutant strains, sequenced and their protein products characterized. Some representative examples are listed in Table 9.3. They are a heterogeneous group, but include proteins with just the properties required for heterochromatin assembly along the lines of the model proposed in Fig. 9.8. One of these genes, designated *Su(var)2-5*, was found to encode the protein HP-1, which one would expect such a screen to detect given the close association of this protein with heterochromatin (see above). Another suppressor, *Su(var)3-7*, encodes a novel zinc finger protein in which the DNA-binding fingers are separated by regions con-

Table 9.3 Examples of suppressors and enhancers of position effect variegation.

Type	Gene/mutant	Protein product
Dominant suppressors of PEV	Su(var)2-5	HP1; associates with heterochromatin and, directly or indirectly, with other proteins
	Su(var)3-7	SU(VAR)3-7; associates with heterochromatin and HP1, seven zinc fingers
	Su(var)3-9	SET domain; histone methyltransferase
	k43	ORC2 (component of replication complex); associates with HP1
	modulo	MODULO; binds DNA and RNA; basic N-terminal and C-terminal domains
Dominant enhancers of PEV	E(var)3-95E	E2F; transcriptional activator and cell cycle regulator
	E(var)62/trithorax-like	GAGA factor; associates with euchromatin; POZ domain
	E(var)3-64BC	RPD3; histone deacetylase
	hel	HEL; ATP-dependent RNA helicase

Data taken from Wallrath (1998).

taining stretches of acidic amino acids. The condensation of chromatin by Su(var)3-7 may therefore involve binding of DNA by the widely separated zinc fingers and interaction of the intervening polyacidic stretches with positively charged chromatin proteins. These proteins are most likely to be histones, particularly as the polyacidic stretches found in Su(var)3-7 resemble those found in histone-binding proteins such as HMG 1 and 2 and nucleoplasmin. However, there are some puzzles as well. Why should the RPD3 deacetylase be an *enhancer* of PEV? All the evidence shows that histones in heterochromatin are under-acetylated. Loss of a deacetylase would be expected to *increase* histone acetylation and therefore reduce (*suppress*) silencing.

How might heterochromatin spread?

Spreading models propose that heterochromatin spreads from its normal location so as to engulf the adjacent gene. What spreads is presumably a complex built up of many interacting proteins that engulfs the DNA and prevents transcription, much as proposed by Zuckerkandl many years ago. Some of these proteins presumably bind DNA and the composition of the euchromatic DNA may play a part in determining the extent to which the complex spreads.

The continuous spreading model makes a clear and testable prediction. If two genes are translocated so as to be adjacent to heterochromatin and if the more distal gene is inactivated, then the more proximal gene must also be inactivated. This is illustrated in Fig. 9.10. Although conceptually simple, the necessary experiments are not easy to carry out in practice, partly because of the lack of appropriate marker genes. Early experimental results were consistent with the predictions of the continuous spreading model, but more recent testing suggests that it is far from the complete story. A recent thorough analysis of the problem has used several different fly lines in which the *white* (*w*) and *roughest* (*rst*) genes are translocated adjacent to heterochromatin. *Rst* expression is monitored by a characteristic eye phenotype that can be visualized microscopically along with the presence or absence of pigment. Frequent examples were found in which *rst*, which was further from the heterochromatin, was *silenced* while *w*, which was closer, was expressed.

In cases where the distal gene is silenced while the proximal one is not, the latter must somehow be protected from the effects of heterochromatin. This suggests that *local* chromatin effects may be able to overcome the influence of heterochromatin and that the final result (i.e. an active or inactive gene) represents the outcome of competition between the two effects. The importance of local effects has been demonstrated by the finding that the frequency with which transgenes inserted next to heterochromatin are silenced can be influenced dramatically by their orientation. It is also interesting to note that transgenes that are located near heterochromatin are more readily switched off when they are inserted as multiple tandem copies than when they are single copy. Repetitive DNA seems to favour silencing. In fact, transgenes inserted anywhere in the genome as tandem arrays are liable to themselves precipitate formation of a heterochromatin-like structure with suppression of transcriptional activity. The various repetitive elements that characterize heterochromatin and that may play a role in its structure and formation are largely absent from euchromatic DNA, and are unlikely to play a role in spreading. However, other DNA elements, as yet unidentified, may promote or inhibit the formation of heterochromatin-like structures. The frequency of such elements will determine whether or not a gene succumbs to the effects of adjacent heterochromatin. In this context, it is interesting that the satellite-DNA-binding protein Prod is also found in euchromatin, possibly as a result of binding to euchromatic DNA elements.

Intranuclear positioning and heterochromatin formation

Positioning models of PEV operate on a completely different premise. They suggest that heterochromatin occupies a particular, defined position in the cell nucleus in which conditions are conducive for compaction of DNA into the structure we define as heterochromatin. (It is not the case that these locations are unsuitable for transcription in general, as evidenced by the presence of some transcriptionally active genes in heterochromatin.) These regions may contain relatively high levels of proteins necessary for heterochromatin assembly. This idea receives strong support from studies in yeast, which show that regions adjacent to the nuclear envelope both locate the telomeres of yeast chromosomes and are rich in the proteins required for assembly of heterochromatin-like structures and transcriptional silencing. This is discussed in Chapter 10. In this model, the translocated gene would still undergo changes in chromatin structure, but there would be no need for spreading of the altered structure along the intervening DNA (Fig. 9.10).

Support for the positioning model of PEV has come from studies of a *Drosophila* mutant called brown dominant (*bw^D*). In this mutation about 2 Mb of satellite DNA has been inserted into the brown gene locus (Fig. 9.11). The initially surprising observation was that this mutation silences not only the *brown* gene containing the insert, but also, in a variegated manner, its normal homologue. This is an example of *trans*-inactivation. This *trans* effect cannot be explained by spreading of heterochromatin along the chromosome from the inserted satellite DNA. A likely explanation has come from the observation that at least some genes in *Drosophila* show somatic pairing, i.e. homologues are located in the same region of the interphase nucleus, and are possibly physically associated. This has been demonstrated in diploid nuclei, for *bw* and other loci, by sophisticated *in situ* hybridization and image analysis. The existence of some sort of pairing mechanism in somatic cells can also be inferred from the exact alignment of gene loci that occurs in polytene chromosomes. The mechanism responsible for somatic pairing remains unknown. Perhaps it uses components required for the well-studied pairing of homologous chromosomes that occurs

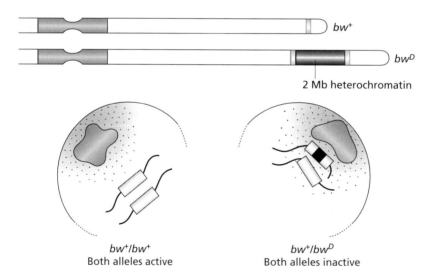

Fig. 9.11 The brown dominant mutation (*bw^D*). In this mutation, the brown (*bw*) gene on chromosome 2 in *Drosophila melanogaster* has been inactivated by insertion of a block of heterochromatin. Surprisingly, the mutation leads to inactivation of the normal *bw* allele on the homologous chromosome. This may be because the mutant gene relocates to a heterochromatin domain in the nucleus and drags the normal allele along with it. Homologous chromosomes are known to have a tendency to pair in the interphase nucleus. The normal allele will be inactivated through proximity to heterochromatin, a form of position effect variegation.

in meiotic cells. But whatever the mechanism, somatic pairing provides an explanation for the bw^D effect. It is proposed that the normal bw homologue is drawn, by somatic pairing, into a region adjacent to the mutated, heterochromatin-containing allele. This region provides an environment that is favourable for heterochromatin formation, but is not compatible with transcription of euchromatin genes. Thus, the normal allele may be silenced. To account for the variegating aspect of the phenomenon, we must suppose that the juxtaposition of the two alleles varies from one nucleus to another. When closely associated, the normal allele is silenced; when further away, it is not. This explanation is outlined in Fig. 9.11.

At this stage, it is probably useful to restate one of the defining characteristics of PEV, namely that the pattern of expression of the variegating allele, once set (i.e. once it is on or off), remains *stable* through many cell generations. If the expression is determined, at least in part, by intranuclear location, as the evidence suggests, then this location must be maintained stably. This could be seen as an example of epigenetic regulation operating at the level of the whole cell nucleus.

Heterochromatin and gene expression in mammals

Most studies of PEV have been carried out in *Drosophila*, but recent work has made it clear that the phenomenon also exists in mammals. This was initially shown by studies using the human *CD2* gene. The gene encodes a surface protein expressed in lymphocytes at particular stages in their development. It is normally under the control of a locus control region (LCR; see Chapter 6), and transgenes containing this LCR are expressed in a position-independent manner in transgenic mice. In contrast, transgenes containing only the 3′ *CD2* enhancer showed variable expression from one mouse line to another. This is how transgenes lacking LCRs usually behave. However, because the *CD2* gene encodes a surface marker, it was possible to use antibody staining to study expression on a cell-by-cell basis. This revealed that, in some lines, expression was variegated, i.e. some cells expressed the gene strongly, while others did not express it at all. Crucially, by using fluorescently labelled DNA probes to locate the *CD2* transgene in metaphase chromosomes, it was shown that the variegating lines had all integrated the transgene in, or adjacent to, centric heterochromatin.

Perhaps the most intriguing parallel between the functional effects of heterochromatin in mammals and flies is the observation that there may be a mammalian equivalent of the brown dominant effect. By *in situ* hybridization with fluorescently labelled DNA probes it has been possible to study the intranuclear location of a set of genes in lymphocytes as the cells move from quiescence to growth. In quiescent cells, genes were distributed throughout the nucleus. When the cells were induced to grow in culture by provision of the appropriate growth factors, genes that were not expressed in lymphocytes moved to posi-

tions adjacent to clumps of centric heterochromatin. Within 3 days of stimulation, one or both alleles of nonexpessed genes were found adjacent to centric
heterochromatin in most cells. This was true both of genes expressed only at
certain points during lymphoid differentiation (such as the surface marker
CD8α and the recombinase *Rag*) and of genes expressed only in other cell
lineages (such as the neuronal-specific gene *Sox1*). It may be that relocation
to a region adjacent to heterochromatin is a general method for gene
silencing in mammals. Intriguingly, it seems that this mechanism may not
operate in transformed cells. A cell line derived from a thymic lymphoma can
be induced to silence *Rag* and other genes by drug treatment or by treatment
with antibodies to its surface T-cell receptor. This silencing was shown to
occur in the absence of any detectable relocation of the genes involved. However, when the cells were examined after several days in culture, genes such as
Rag were found to be re-expressed. This result suggests that relocation is not
necessary for the initial suppression of transcription, but is required for the
long-term maintenance of the transcriptionally inactive state. Whether this
process involves the heterochromatin proteins or mammalian Su(var)s remains to be seen, but a lymphoid-specific transcription factor named Ikaros undergoes a parallel relocation to heterochromatin clusters in growth-stimulated
lymphocytes and is likely to play an important role in the process, possibly
through its association with histone deacetylases (Chapter 8). The story is summarized in Fig. 9.12.

A catalytically active mammalian Su(var)

Humans and mice have homologues of *Drosophila* Su(var)s that seem to play
very similar roles to their fly counterparts. For example, human and mouse versions of Su(var)3-9 locate to heterochromatin foci and interact with the mammalian homologues of HP-1. Also a single copy of a transgene encoding the
human homologue, SUV39H1, when introduced into the appropriate fly
strains, was able to enhance PEV of the translocated *white* gene just as well as the

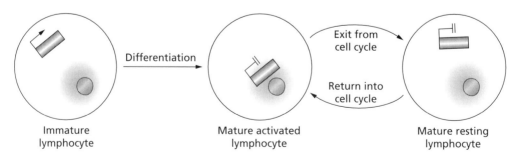

Fig. 9.12 Certain genes change position, relative to centromeric heterochromatin domains
(shaded), during lymphocyte differentiation. (Adapted from Brown *et al.*, 1999.) Note that
the peri-centromeric positioning is relaxed when mature cells become quiescent and exit
from the cell cycle.

Drosophila version. (Remember, Su(var)s *enhance* PEV, i.e. *increase* silencing, when present in three or more copies and *suppress* it when present in only a single copy.)

A major step forward in understanding how Su(var)s might work has come with the finding that both the human and mouse homologues of Su(var)3-9 (SUV39H1 and Suv39h1, respectively) have histone methyltransferase activity (see Fig. 4.5 for details of this reaction). Other histones or other H3 lysines were not detectably methylated. The catalytic site is located within the SET domain, a conserved region found in many proteins, including the *PcG* protein E(Z) and the *trithorax* group protein TRX (Chapter 10). However, no other SET domain protein has so far been shown to have histone methyltransferase activity and it seems likely that the activity of SUV39H1 depends on both the SET domain and residues in adjacent, cysteine-rich regions. The domain structure of SUV39H1 is shown in Fig. 9.13.

In *in vitro* assays, recombinant Suvar39h methylates H3 specifically at lysine 9. The significance of this is, as noted earlier, that the heterochromatin protein HP1 binds specifically, via its chromodomain, to H3 methylated at this site. So is H3me9, generated by the action of Suvar39h, the signal that targets HP1 to specific chromatin domains and leads to heterochromatin formation? It seems that this may be only part of the answer. While Suvar39h associates *preferentially* (not exclusively) with heterochromatin *in vivo*, H3me9 does not. In fact, it seems that only a small proportion of the cell's total H3me9 is generated by the action of Suvar39h. (Mice lacking the enzyme have near normal levels of H3me9 in bulk histones.) Perhaps HP1 requires *both* methylation of lysine 9 and low levels of histone acetylation (another almost universal characteristic of heterochromatin) before it can bind to chromatin. The question also remains as to how Suvar39h might be targeted to heterochromatin domains in the first place. Perhaps a heterochromatin-binding protein is involved. There is a precedent for this, in that both Suvar39h and HP1 have recently been shown to associate with the transcriptional repressor protein Rb, which targets them to selected promoters. This leads to the important conclusion that HP1, Suvar39h and (presumably) H3me9, have roles in transcriptional control over and above their involvement in heterochromatin assembly.

Both mice and humans each have two close homologues of Su(var)3-9. How these differ in function or catalytic activities remains to be defined. Mice in which either homologue is knocked out are viable, but double-knockout mice are recovered from the relevant crosses at only about 25% of the expected frequency and the survivors show retarded growth. Significantly, embryonic

Fig. 9.13 The domain structure of SUV39H1. The SET domain lies between residues 249 and 375. Cysteine-rich regions are lightly shaded.

fibroblasts established from the double-knockouts showed increased numbers of cells with abnormal nuclear morphology consistent with defects in chromosome segregation at mitosis. They also had increased levels of H3 phosphorylated at serine 10. The importance of this observation rests on the fact that H3S10 phosphorylation is necessary for the normal progression of cells through mitosis. As noted earlier, the modification is first detected in late G2 in centric heterochromatin and then spreads across the entire chromosome complement. But why does absence of Suv39h1/2 alter H3 phosphorylation? A possible answer comes from the results of *in vitro* assays of the H3 kinase responsible for H3S10 phosphorylation (IpI1/aurora). H3 methylated at lysine 9 is a much less efficient substrate for this enzyme than unmethylated H3. Perhaps methylation of H3 at lysine 9 is used to regulate phosphorylation of H3 at serine 10 in heterochromatin as cells prepare for mitosis. Much remains to be done, but these findings not only provide an intriguing insight into the chain of events that might determine Su(var) action, but also demonstrate very clearly how different modifications to the histone tails might interact to exert specific functional effects.

Further Reading

General reviews

Hennig, W. (1999) Heterochromatin. *Chromosoma*, **108**: 1–9.
Strachan, T. & Read, A. P. (1997) *Human Molecular Genetics*. Bios Scientific Publishers, Oxford, pp. 183–209.
Wallrath, L. L. (1998) Unfolding the mysteries of heterochromatin. *Curr. Opin. Genet. Dev.*, **8**: 147–153.

Heterochromatin DNA

Gatti, M. & Pimpinelli, S. (1992) Functional elements in *Drosophila melanogaster* heterochromatin. *Ann. Rev. Genet.*, **26**: 239–275.
Lohe, A. R., Hilliker, A. J. & Roberts, P. A. (1993) Mapping simple repeated DNA sequences in heterochromatin of *Drosophila melanogaster*. *Genetics*, **134**: 1149–1174.
Pimpinelli, S., Berloco, M., Fanti, L. *et al.* (1995) Transposable elements are stable structural components of *Drosophila melanogaster* heterochromatin. *Proc. Natl. Acad. Sci., USA*, **92**: 3804–3808.

Heterochromatin genes

Weiler, K. S. & Wakimoto, B. (1995) Heterochromatin and gene expression in *Drosophila. Ann. Rev. Genet.*, **29**: 577–605.

Heterochromatin proteins

Aagaard, L., Laibl, G., Selenko, P. *et al.* (1999) Functional mammalian homologues of the *Drosophila* PEV modifier *Su(var)3–9* encode centromere-associated proteins which complex with the heterochromatin component M31. *EMBO J.*, **18**: 1923–1938.

Bannister, A. J., Zegerman, P., Partridge, J. F. *et al.* (2001) Selective recognition of methylated lysine 9 on histone H3 by the HP1 chromodomain. *Nature*, **410:** 120–124.

Eissenberg, J. C. & Elgin, S. C. R. (2000) The HP1 protein family: getting a grip on chromatin. *Curr. Opin. Genet. Dev.*, **10:** 204–210.

Lachner, M., O'Carroll, D., Rea, S., Mechtler, K. & Jenuwein, T. (2001) Methylation of histone H3 lysine 9 creates a binding site for HP1 proteins. *Nature*, **410:** 116–120.

Nielsen, S. J., Schneider, R., Bauer, U.-M. *et al.* (2001) Rb targets histone H3 methylation and HP1 to promoters. *Nature*, **412:** 561–565.

Raff, J. W., Kellum, R. & Alberts, B. (1994) The *Drosophila* GAGA transcription factor is associated with specific regions of heterochromatin throughout the cell cycle. *EMBO J.*, **13:** 5977–5983.

Rea, S., Eisenhaber, F., O'Carroll, D. *et al.* (2000) Regulation of chromatin structure by site-specific histone H3 methyltransferases. *Nature*, **406:** 593–599.

Török, T., Gorjánácz, M., Bryant, P. & Kiss, I. (2000) Prod is a novel DNA-binding protein that binds to the $1.686\,g/cm^3$ 10 bp satellite repeat of *Drosophila melanogaster*. *Nucl. Acids Res.*, **28:** 3551–3557.

Heterochromatin and transcription—position effect variegation

Belyaeva, E. S., Koryakov, D. E., Pokholkova, G. V., Demakova, O. V. & Zhimulev, I. F. (1997) Cytological study of the Brown Dominant position effect. *Chromosoma*, **106:** 124–132.

Festenstein, R., Tolaini, M., Corbella, P. *et al.* (1996) Locus-control region function and heterochromatin-induced position effect variegation. *Science*, **271:** 1123–1125.

Hennikoff, S. (1994) A reconsideration of the mechanism of position effect. *Genetics*, **138:** 1–5.

Sabl, J. F. & Hennikoff, S. (1996) Copy number and orientation determine the susceptibility of a gene to silencing by nearby heterochromatin in *Drosophila*. *Genetics*, **142:** 147–158.

Talbert, P. B. & Hennikoff, S. (2000) A reexamination of spreading of position effect variegation in the *white-roughest* region of *Drosophila melanogaster*. *Genetics*, **154:** 259–272.

Zhimulev, I. F., Belyaeva, E. S., Fomina, O. V., Protopopov, M. O. & Bolshakov, V. N. (1986) Cytogenetic and molecular aspects of position effect variegation in *Drosophila melanogaster*. *Chromosoma*, **94:** 492–504.

Intranuclear positioning

Brown, K. E., Baxter, J., Graf, D., Merkenschlager, M. & Fisher, A. G. (1999) Dynamic repositioning of genes in the nucleus of lymphocytes preparing for cell division. *Mol. Cell*, **3:** 207–217.

Hennikoff, S., Jackson, J. M. & Talbert, P. B. (1995) Distance and pairing effects on the Brown (Dominant) heterochromatin element in *Drosophila*. *Genetics*, **140:** 1007–1017.

Pyrpasopoulou, A., Meier, J., Maison, C., Simos, G. & Georgatos, S. D. (1996) The lamin B receptor (LBR) provides essential chromatin docking sites at the nuclear envelope. *EMBO J.*, **15:** 7108–7119.

Chapter 10: Long-term Silencing of Gene Expression

Introduction

In a typical human nucleus, the great majority of genes will be switched off. The reason is that most genes are required only in particular cell types and are expressed only in those cells in which they are needed. In a typical human cell, as many as 20 000–30 000 genes may be in this 'not needed' category. Given the large numbers involved, it is important that transcription of these genes is not just down-regulated, but is completely switched off, a process that is generally referred to as *silencing*. Even a small amount of transcription from many thousands of genes would be sufficient to disrupt the cell's control systems, as well as making unmanageable demands on the transcription machinery. A mechanism for silencing genes was probably necessary to allow evolution of the large genomes characteristic of multicellular eukaryotes. However, silencing is not limited to higher eukaryotes. In single-celled eukaryotes such as yeast, in which only a few genes are ever silenced, silencing can be crucially important in situations where even a small amount of the gene product would disrupt the cell's behaviour.

There are several different mechanisms that eukaryotes can adopt to silence their genes. The combination of mechanisms used varies between organisms and between genes, but most bring about changes in chromatin structure, in one way or another. For mammals, DNA methylation provides an important starting point both for initiating the changes in chromatin structure that lead to silencing and for maintaining the silent state.

DNA methylation

Cytosine residues in DNA can be methylated *in vivo* by the enzyme DNA methyltransferase (Dnmt). Two aspects of the specificity of Dnmt are important. First, it will *only* methylate cytosines that are part of CpG dimers; other cytosines are not recognized. This may be attributable to the fact that CpG forms a symmetrical structure in double-stranded DNA, i.e. a CpG dimer, in the 5′–3′ direction, occurs on both strands (Fig. 10.1). Second, the enzyme shows a preference for pairs of CpG dimers in which one of the two cytosines is already methylated (Fig. 10.1). This preference has important implications for the maintenance of patterns of methylation through DNA replication.

It has been known for more than 40 years that CpG dimers are much rarer than one might expect in mammalian DNA. For example, C and G bases in human DNA each occur with a frequency of about 0.2, so the frequency of CpG

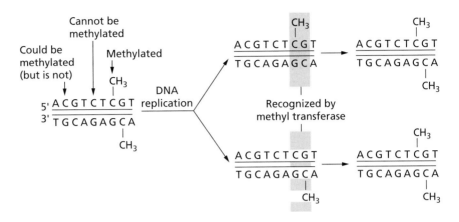

Fig. 10.1 DNA methyltransferase efficiently methylates hemimethylated sites in DNA and maintains patterns of cytosine methylation after DNA replication.

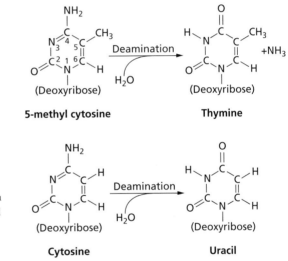

Fig. 10.2 Spontaneous chemical deamination of 5-methyl cytosine forms thymine, a normal constituent of DNA, whereas deamination of cytosine forms uracil, a base normally found only in RNA.

dimers should be $0.2 \times 0.2 = 0.04$. In fact it is 0.008, five times less than it should be. Of those CpGs that *are* present, most (60–90%, depending on the organism) are methylated. This high frequency of methylation provides a likely explanation for the dearth of CpG dimers. The explanation is based on the fact that cytosine residues are susceptible to deamination by spontaneous hydrolysis, resulting in conversion to uracil (Fig. 10.2). This is a chemical event that happens, as a matter of course, at a rate estimated to be about 100 bases per (mammalian) genome per day. In normal circumstances, the uracil is recognized as an abnormal base (for DNA) by the DNA repair mechanism, which brings about its excision and replacement. However, if 5-methyl cytosine (5-MeC) undergoes the same deamination, the resulting base is not uracil but thymine (Fig. 10.2), a normal constituent of DNA with a high probability of escaping the attentions of the repair systems and becoming fixed. The low frequency of CpG dimers in many vertebrates probably reflects their progressive loss over evolutionary time through this C→T chemical conversion process (see below).

Of the CpGs that remain, most are scattered through the genome, but some are clustered in regions of relatively high CpG density known as 'CpG islands'. These are usually between 500 and 200 bp in size and often incorporate the promoter regions of genes. A typical CpG island is shown in Fig. 10.3. Note that the frequency of CpGs in the island is comparable with that of GpCs and is about what one would expect based on the G+C frequency. This implies that CpG islands are regions that have escaped the loss of CpGs during evolution that has occurred in the rest of the genome. The reason for this may lie in the finding that whereas CpGs scattered through the genome are usually methylated, those in CpG islands are not. This lack of methylation means that they are not particularly susceptible to loss through chemical conversion. This is illustrated in Fig. 10.4. Note that the proposed mechanism does not require that CpG islands serve any useful purpose or confer any evolutionary advantage. Indeed, it is quite possible that CpG islands are an evolutionary by-product. Inaccessibility of certain (promoter-proximal) CpGs to DNA methyltransferase, combined with the genome-wide loss of meC through mutation, would give exactly the type of CpG distribution that we see today.

CpG base pairs across the mouse *Aprt* gene

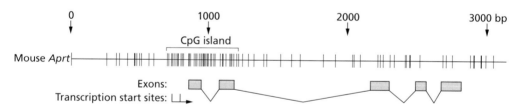

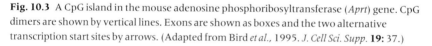

Fig. 10.3 A CpG island in the mouse adenosine phosphoribosyltransferase (*Aprt*) gene. CpG dimers are shown by vertical lines. Exons are shown as boxes and the two alternative transcription start sites by arrows. (Adapted from Bird *et al.*, 1995. *J. Cell Sci. Supp.* **19**: 37.)

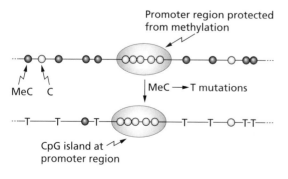

Fig. 10.4 Progressive conversion of 5-methyl cytosine residues to thymine through evolution results in formation of a CpG island at regions protected from methylation.

What is the purpose of DNA methylation? Some organisms (e.g. yeast, fruit-flies and fungi) have little or no DNA methylation, so it is clearly not required for fundamental life processes. On the other hand, mutant mice with severely reduced levels of DNA methyltransferase activity die early in development. The DNA methylation mechanism, once in place, is clearly essential. But what was the evolutionary advantage that led to the establishment of a methylation system in the first place? It must have been substantial, given that the mutability of 5-MeC (Fig. 10.2) will necessarily place a selective burden on the organism. A likely explanation is based on experimental evidence, discussed in the next section, which shows that CpG methylation is closely associated with suppression of transcription and long-term gene silencing. The importance of efficient silencing mechanisms for the large genomes of multicellular organisms has already been emphasized. It has been suggested that the silencing property of methylation provided an additional advantage by suppressing the transcription of retroviruses and transposable elements, thereby inhibiting their ability to spread throughout the genome via their RNA intermediates. The high frequency of transposable elements in higher eukaryotes has been noted previously (Chapter 9) and mechanisms to limit their spread, with its consequent genomic damage, would be a major benefit.

Methylation and gene silencing

An early link between DNA methylation and transcriptional inactivity was made by using antibodies to methyl cytosine to stain metaphase chromosome spreads. These immunolabelling experiments showed that the highest levels of methylation were found in blocks of centromeric heterochromatin. The ability of methylation to lower transcription from individual genes was demonstrated by experiments measuring transcription from *in-vitro*-methylated and nonmethylated constructs transiently transfected into tissue culture cells. Transcription from the methylated construct was always weak or absent. An important insight was gained by experiments in which transcription levels were measured over time following injection of methylated and nonmethylated constructs into *Xenopus* oocytes. Initially, levels of transcription were comparable, but after several hours suppression of the methylated construct became apparent. One interpretation is that silencing occurs only after the injected DNA has been assembled into chromatin, or at least complexed with endogenous proteins. This is consistent with the other results showing that methylation has little or no effect on rates of transcription *in vitro* from naked DNA templates and that methylated DNA integrated into the genome exhibits a distinctive chromatin structure.

Proteins that bind methylated DNA

An important clue about the mechanism by which cytosine methylation can

suppress transcription came from the identification of proteins that bound preferentially to methylated DNA. The first of these to be characterized, designated
MeCP1, is a large protein complex (400–800 kDa) that binds to DNA containing
a stretch of at least 12 symmetrically methylated CpGs (CpG dimers in which the
C residues are methylated on *both* strands; the usual situation *in vivo*). The
DNA-binding subunit of MeCp1 has now been identified as a 44 kDa protein
designated MBD2. It was shown that MeCP1 could repress transcription from
methylated promoters *in vitro* and that the extent of repression was dependent
on both the density of methylation and the strength of the promoter. Thus, a
weak promoter could be inhibited by a low level of methylation, whereas a *strong*
promoter (strengthened by an enhancer, for example) was inhibited only by
heavy methylation, which would give a high affinity for MeCp1. This is illustrated in Fig. 10.5. It is important that this regulatory mechanism depends on the
density of methylated CpGs across a chromosome domain, rather than on
whether individual CpG dimers are methylated or not.

Not all methyl-DNA-binding proteins require runs of methylated CpGs
for binding. A second meDNA-binding protein, MeCP2, consists of a single
polypeptide chain of about 52 kDa and is capable of binding to a single symmetrically methylated CpG dimer. It has abundant binding sites in chromatin, can
incorporate itself into preassembled (methylated) chromatin with displacement of H1 and has been shown microscopically to be concentrated in heterochromatic regions known to be rich in methyl-CpG. Transient transfection
experiments in cultured cells show MeCP2 to be an effective transcriptional
repressor. Mice lacking the *Mecp2* gene develop severe neurological defects
early in adult life. MeCP2 contains a methyl-CpG-binding domain (MBD) that

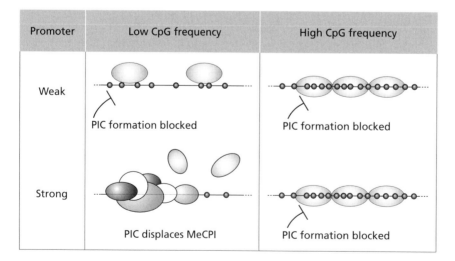

Fig. 10.5 Repression by MeCP1 depends on the frequency of methyl CpG dimers and the
strength of the promoter.

is homologous to the DNA binding region of the MBD2 protein in MeCP1. Four other proteins have so far been identified that contain closely homologous MBD domains, although the *in vivo* roles of these proteins have still to be determined. The members of the MBD protein family are briefly described in Fig. 10.6.

MeCP1 and MeCP2 repress transcription *in vitro* and *in vivo*

Purified or recombinant MeCP2 inhibits *in vitro* transcription from methylated DNA but not from the equivalent nonmethylated DNA (with which it produces slightly elevated transcription). Mutations in the MBD abolish inhibition. Removal of the C-terminal domain was found to reduce inhibition of transcription, suggesting that the inhibition may not be the result simply of steric blocking by binding to promoter proximal DNA, but may require additional, protein–protein interactions. In order to identify this potential repression domain, *in vivo* inhibitory effects were tested by transfection into cultured cells

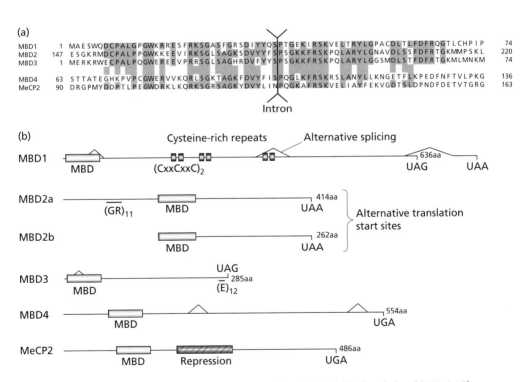

Fig. 10.6 Domain structure of proteins that selectively bind methylated DNA. (a) The methyl-CpG-binding domains (MBDs) of the proteins MBD1, 2, 3, 4 and MeCP2 have a high level of amino acid sequence conservation. Residues shared by three or more of the five proteins are shaded. (b) Domain structure and alternative splice variants of MBD proteins. MBD2a has a glycine.arginine (GR) repeat sequence and MBD3 has a glutamic acid (E) repeat at its C-terminal end. (Adapted from Hendrich & Bird 1998.)

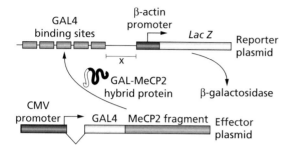

Fig. 10.7 DNA constructs used to test the inhibitory effects of MeCP2 in transient transfection assays. Fragments of MeCP2 that can exert a repressive effect will cause a reduction in β-galactosidase activity in transfected cells. A control construct lacking GAL4 sites should not be affected.

of MeCP2 attached to the *GAL4* DNA binding domain, together with a 'reporter' plasmid in which five GAL4 binding sites were placed upstream of a β-actin promoter driving *LacZ* (Fig. 10.7). This somewhat indirect approach is made necessary by the fact that methylated reporters are shut down rapidly by endogenous mechanisms when transfected into mammalian cells, so cannot be used for this type of experiment. The MeCP2–GAL4 construct *repressed* reporter transcription. By analysing the inhibitory effectiveness of deleted MeCP2 constructs, a minimal transcription repression domain (TRD) was located between amino acids 207 and 310 (Fig. 10.6). This region has a high frequency of basic residues (26% lysine and arginine). Transient transfections were also used to show that MeCP2 binding can effectively repress transcription even when it binds at distances of up to 2 kb from the promoter. This was demonstrated by inserting one or more DNA fragments of about 900 bp between the GAL4 binding sites on the reporter construct and the promoter (region 'x' in Fig. 10.7).

Initial experiments on the inhibitory properties of MeCP1 were hindered by its multi-subunit composition and the difficulty of achieving high levels of purification. The recent identification and cloning of the methyl-CpG-binding component of the MeCP1 complex (MBD2) has allowed transient transfection approaches to be applied to this protein. MBD2 is able to repress transcription both from methylated reporter constructs in HeLa cells (in which endogenous MeCP2 is present at only extremely low levels) and, when linked to GAL4, from reporters containing GAL4 binding sites (i.e. the approach outlined in Fig. 10.7).

MBD proteins are associated with HDAC-containing repressor complexes

Quantitative Western blotting has shown that MeCP2 is a relatively rare molecule. There are about 10^6 molecules of MeCP2 per nucleus in a typical mammalian cell (6×10^6 in brain), compared with (typically) 4×10^7 methyl-CpGs. So, most MeCpGs are free of MeCP2. However, it is clear from the results of *in vitro* transcription and transient transfection assays that binding of only a few

MeCP2 molecules can spread repression across a relatively large domain. How is this achieved?

Both MeCP1 (MBD2) and MeCP2 are now known to be associated with multiprotein complexes containing the histone deacetylases HDAC1 and HDAC2. Immunoprecipitation with antibodies to MeCP2 or MBD2 brings down histone deacetylase activity in the precipitate, in which HDAC1 and HDAC2 are also identified by Western blotting (along with various other proteins). The functional significance of the deacetylase connection has been confirmed by showing that the inhibition of transcription from reporter constructs by cotransfected MeCP2 or MBD2 is sensitive to the HDAC inhibitor Trichostatin A (TSA). These enzyme complexes are important components of the cell's chromatin-modifying machinery and were discussed in detail in Chapter 8. Proteins such as MeCP2 and the HDAC-containing complexes with which they associate provide a link between DNA methylation, modification of chromatin structure and, consequently, regulation of transcription. A model summarizing these links is shown in Fig. 10.8.

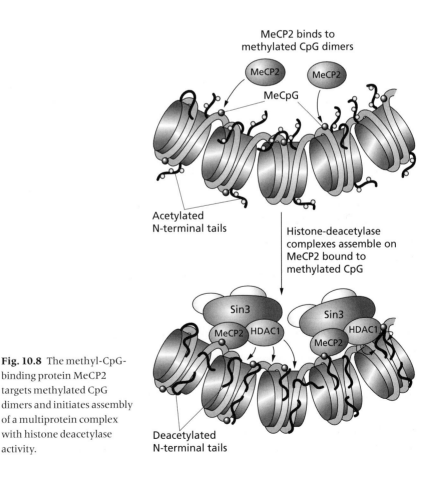

Fig. 10.8 The methyl-CpG-binding protein MeCP2 targets methylated CpG dimers and initiates assembly of a multiprotein complex with histone deacetylase activity.

As always, caveats must be applied to this attractive model. First, there are certainly protein components of the inhibitory complex shown in the figure that we don't yet know about. Indeed, there may be several different types of complex. Second, it is important to remember that histones are not the only protein components of the transcription initiation complex that can be acetylated and deacetylated (see Chapter 8). Histones themselves may not be, functionally, the most important targets of the deacetylases. In considering the role of the histones, it is also important to remember that *transiently transfected* constructs do not adopt a normal chromatin structure (i.e. they do not assemble canonical nucleosomes). Effects observed with such constructs may not be exactly analogous to those that occur on endogenous genes. In *in vitro* transcription assays, using naked methylated or unmethylated constructs, histones may not feature at all. Third, there is evidence for a repression mechanism mediated by a component of HDAC co-repressor complexes (Sin3a) that is *independent* of HDAC activity and may involve direct interaction with the basal transcription complex. Such a mechanism would help explain why inhibition of deacetylase activity effectively suppresses MBD-mediated inhibition from some promoters but not others. It may be that the HDACs can, on occasion, play a structural role in assembling an inhibitory complex that does not require their catalytic activity. It is also interesting to note that an MBD protein has been detected in *Drosophila*, an organism in which DNA methylation is minimal. This protein, dMBD2/3, does not bind preferentially to methylated DNA *in vitro*, but does bind to dHDAC1 both *in vivo* and *in vitro*.

Demethylation of DNA

DNA methylation is a chemically stable modification of genomic DNA that can constitute a long-term epigenetic marker specifying (usually) transcriptional inactivity. However, in some situations patterns of methylation can be remarkably dynamic. For example, in very early embryos there is a dramatic transition from the overall high frequency of CpG methylation characteristic of germ cells to the low frequency characteristic of the earliest stages of embryogenesis. Some mechanism must exist for bringing this about, but what it is remains something of a mystery. At first sight, removal of the methyl group from cytosine might seem a fairly straightforward task for the appropriate enzyme. However, demethylation presents a chemical problem, in that it requires the cleavage of a C–C bond, an event that demands an extremely high activation energy and of which there are very few examples in living organisms. It has recently been proposed that the protein MBD2 (or at least a protein closely associated with it) has just such a demethylating capability, but the matter remains controversial. An alternative mechanism, for which the biochemistry is more secure, requires that the whole methyl cytosine base is removed by a DNA glycosylase, leaving the deoxyribose backbone intact. The gap is then filled with a nonmethylated nucleotide by the DNA repair system (Fig. 10.9). It is still not

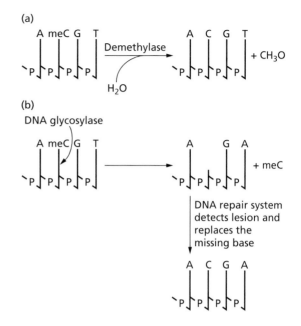

Fig. 10.9 Two possible mechanisms by which DNA can be demethylated. (a) It has been proposed that a demethylating enzyme might cleave the bond between the 5′ carbon of cytosine and the attached methyl group (see Fig. 10.2). The activation energy needed to cleave such C–C bonds is exceptionally high and there are very few examples of enzymes capable of carrying out such a reaction. (b) An alternative demethylating pathway could involve removal of the whole methyl-C base from its associated ribose moiety by a DNA glycosylase. Enzymes of this type are well known. The DNA lesion so produced can then be repaired by enzymes of the DNA repair pathway.

clear whether this mechanism provides a major pathway for rapid demethylation *in vivo*.

An alternative mechanism that provides less rapid demethylation is to prevent methylation of hemi-methylated CpG dimer pairs following DNA replication in S-phase. The Dnmt1 enzyme has a preference for such dimers and in the normal course of events they will be rapidly brought up to the fully methylated state (Fig. 10.1). If newly replicated CpGs were somehow sheltered from Dnmt1, there would be a progressive diminution in methylation levels with each succeeding round of DNA replication, with no need for active demethylation.

Silencing at telomeres and mating type loci in yeast

The advantages of yeast as an experimental eukaryote are numerous and have been noted previously. It allows use of a combination of biochemical and genetic approaches in order to unravel complex molecular mechanisms. If chromatin from either of the commonly used laboratory yeasts (*Saccharomyces cerevisiae* and *S. pombe*) is artificially spread and examined under the electron microscope, most of it is in the form of 10 nm beads-on-a-string fibres. There is little sign of the higher-order chromatin structures characteristic of higher eukaryotes. It seems that most of the yeast genome is in a relatively extended configuration, a conclusion supported by the finding that it is almost uniformly sensitive to nuclease digestion. These complementary results suggest that yeast is unlikely to contain very much in the way of heterochromatin, and indeed nothing resembling heterochromatin can be identified in yeast by conventional microscopical analysis. However, despite this, there are three examples in yeast of gene silencing through chromatin structures that are reminiscent of the behaviour of heterochromatin in higher eukaryotes. These are the silencing of genes at telomeres and at the mating type loci in *S. cerevisiae* and the silencing of genes adjacent to centromeres in *S. pombe*. The first two will be considered here and the third will be deferred until we consider epigenetics and cell memory in the next chapter.

Telomeric silencing

The structure of the telomeres of *S. cerevisiae* is shown in Fig. 10.10. They are characterized by terminal repeats of the sequence $C_{1-3}A$ (about 350 bp in all) followed by 10–15 kb of middle repetitive elements. In *S. cerevisiae*, the telomeric repeats are packaged in a non-nucleosomal stucture called the telosome. Nucleosomes are found along the middle repetitive elements and a conventional chromatin structure is established well before the first coding regions are encountered.

The primary function of the telomere in most, if not all, eukaryotes is to protect the end of the chromosome from the progressive loss of DNA that would, in its absence, inevitably occur during each round of DNA replication. The DNA replication mechanism is such that when the polymerase complex reaches a chromosome end, 20 or so bases at the 3′ end of the original DNA strand are left unreplicated. If this were allowed to continue unchecked, the unpaired DNA would be lost and the chromosomes would become progressively shorter. In order to prevent this, an enzyme called telomerase is used to replicate the stretches of single-stranded DNA at the chromosome ends. The $C_{1-3}A$ repeats are essential for recognition by telomerase. Very similar mechanisms are used to prevent telomere shortening in all eukaryotes, although the nature of the telomeric repeat sequence varies, as does the exact structure of the telomere itself.

Telomeres clearly have a vital role in maintaining the integrity of chromosomes, but they have another ability that has given them an important place in studies of gene regulation, namely the ability to suppress transcription from genes that are positioned adjacent to them.

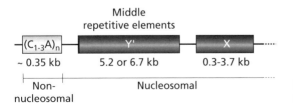

Fig. 10.10 Telomere structure in *S. cerevisiae*.

A particularly graphic way of monitoring this is through use of the *ADE2* gene. *ADE2* encodes an enzyme that prevents formation of a red pigment by certain yeast strains, with the result that colonies of *ADE2+* cells are white and colonies of *ADE2−* cells are red. If yeast strains are constructed in which the *ADE2* gene is repositioned adjacent to a telomere, then it is often silenced, with the result that cells that were originally white turn red (Fig. 10.11). However, the silencing is not seen in all cells and these strains, when plated out, will display a mixture of red (*ADE−*) and white (*ADE+*) colonies. The silencing is relatively stable from one cell generation to the next, but occasional switching during colony growth gives rise to mixed, 'sectored' colonies. This situation is closely analogous to the variegated expression of genes translocated to positions adjacent to centric heterochromatin in *Drosophila*. The extent of variegation (i.e. silencing) can be monitored by counting red and white colonies. This simple read-out of sometimes complex genetic manipulations was used to demonstrate the existence of position effect variegation (PEV) in yeast. A closely related approach uses a selectable marker gene, *URA3*, instead of *ADE2*. In the presence of 5-fluoro-orotic acid (5-FOA), cells in which *URA3* is silent survive, whereas those in which it is active die, owing to the conversion of 5-FOA to a toxic product (Fig. 10.12).

Using these and related approaches it has been shown that the $C_{1-3}A$ repeats adjacent to a telomere can repress genes within 3–4 kb. The $C_{1-3}A$ repeats are *necessary* for telomeric silencing, but they are not *sufficient*. If blocks of $C_{1-3}A$ repeats are translocated away from the telomere, they do not then repress adjacent genes. Some additional property of the telomere is clearly required for silencing. Before going on to ask what this property might be and explore

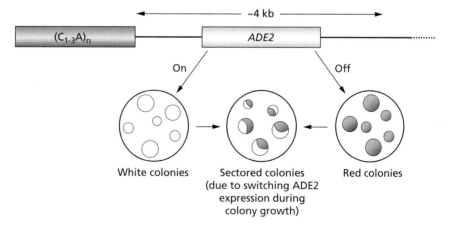

Fig. 10.11 Telomeres can repress transcription of test genes inserted close to them. In the example shown in the diagram the yeast *ADE2* gene has been inserted close to a telomere. The *ADE2* gene product is involved in pigment metabolism. When the gene is on, yeast colonies are white and when it is off they are red. If the gene switches from one state to another during colony growth, the resulting colony will have both red and white sectors.

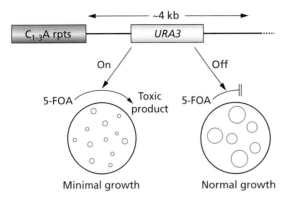

Fig. 10.12 Telomeres can repress transcription of test genes inserted close to them. Expression of the *URA3* gene in yeast can be assayed by the ability of cells to grow in 5-fluoro-orotic acid (5-FOA). The compound itself it harmless, but is converted to a toxic product in cells in which the *URA3* gene is active. Colony growth can be used as an assay of *URA3* activity.

telomeric silencing in more detail, it will be useful to consider the second example of chromatin-mediated gene silencing in yeast, namely silencing of the mating type genes. It has become clear that the two silencing mechanisms share many of the same proteins and are closely interlinked.

The mating type genes and mating type switching

For most of the time, yeasts are haploid and their lives are untroubled by the demands of sexual reproduction. But sometimes, usually when nutrient levels begin to fall, two yeast cells will fuse to form a diploid cell that can then go through meiosis and form haploid spores. These can, if necessary, survive for long periods under adverse environmental conditions and re-enter the growth and division cycle when things improve.

In order to mate successfully, cells of the yeast *S. cerevisiae* must adopt one or other of two mating types, designated **a** or α. This means that they produce mating factors (one or two specific proteins) appropriate to that type and respond to the factors of the opposite type (Fig. 10.13). (A similar system is used by *S. pombe* but the terminology differs.) The **a** and α mating types can be thought of as the yeast equivalent of males and females. The end result of the mating process is a diploid cell in which *both* **a** and α factors are produced. The presence of both types of factor switches off genes necessary for mating, so diploid cells cannot mate, but progress instead through meiosis.

Yeasts have the ability to switch from one mating type to the other. The frequency of switching varies between strains, but can be frequent, up to once every cell cycle. Yeasts can do this because they use an unusual 'cassette' mechanism to determine their mating type. Mating factors are produced by inserting a copy of either the **a**-determining or the α-determining genes at the mating locus, *MAT*. Yeast cells keep copies of both mating type genes at two normally silent loci called *HML*α and *HMR***a**. When switching mating type, the cell

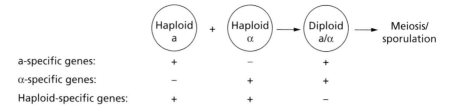

	Haploid a	+	Haploid α	→	Diploid a/α	→	Meiosis/ sporulation
a-specific genes:	+		–		+		
α-specific genes:	–		+		+		
Haploid-specific genes:	+		+		–		

Fig. 10.13 Mating in *S. cerevisiae*.

'chooses' the **a** or α mating type gene and copies it into the *MAT* locus (displacing the incumbent gene), where it is expressed. The process is described in Fig. 10.14. The simple scheme does not do justice to what is a quite extraordinary mechanism involving a precise rearrangement of DNA over large distances. *HMLα* and *HMRa* are separated from MAT locus by about 95 kb and 200 kb of DNA, respectively (Fig. 10.14).

The important points for the present discussion are that (i) the mating type loci *HMLα* and *HMRa* are normally silent, and (ii) if they are expressed, then the cell will contain both **a** and α mating factors, will behave as if it were diploid and will be unable to mate (i.e. will be functionally sterile). It is obviously crucially important (for yeast) that the mating type genes at *HMLα* and *HMRa* remain silenced. Loss of silencing at these loci is detected easily (in laboratories set up to do this sort of thing) by measuring changes in the frequency of mating (*mating efficiency*).

Silencing at *HMLα* and *HMRa* requires two DNA elements designated E and I. The structure of these 'silencers' is shown in Fig. 10.15. The two silencing elements each contain sequences that can act as binding sites for the proteins Rap1p and Abf1p and for the protein complex that initiates DNA replication, the origin recognition complex, ORC (Fig. 10.15). Despite their similarities, there are significant functional differences between the silencers. For example, all silencing elements can act as origins of replication when inserted into plasmids, but only the *HMR*-E and -I elements have detectable activity as *natural*, chromosomal origins. Also, *HMR*-E can function on its own to silence genes inserted close to it, whereas *HMR*-I cannot. Studies of the *HMR*-E silencer have shown that at least two of the three protein binding sites within it must be deleted before it loses its silencing ability (i.e. there is functional redundancy in the silencing mechanism). In order to explore the roles of the three sites further, oligonucleotides modelled on the three *HMR*-E silencing elements have been used to construct an artificial *HMR*-E silencer 142 bp in length. Functionally, this construct can replace an 800 bp region incorporating the endogenous *HMR*-E. However, in the artificial silencer, mutation of any one of the three binding sites within it destroys its silencing function. The redundancy present in the endogenous silencer is lost in the stripped-down, artificial version and must rely

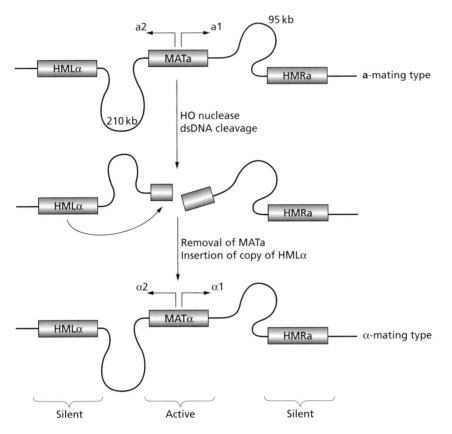

Fig. 10.14 Mating-type switching in *S. cerevisiae*.

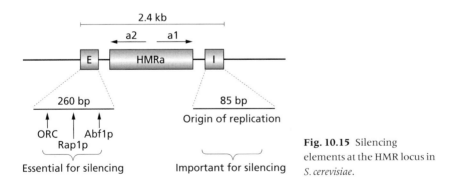

Fig. 10.15 Silencing elements at the HMR locus in *S. cerevisiae*.

on sequences or structure outside the three protein binding sites identified so far.

These results introduce an important point that crops up time and time again when considering the functional effects of chromatin, namely that any particular effect can rarely be attributed to just a single cause. Instead, it is usually a consequence of the cumulative effects of several different parameters, each one of which may be *necessary*, but is rarely, if ever, *sufficient*. These usually include different proteins and DNA sequences, but may also include metabolic effects, state of differentiation or even environmental parameters.

Both telomeres and mating type loci show heterochromatin-like properties

There are fundamental similarities between silencing at telomeres and at mating type loci. In the case of the mating type genes, two identical genes (in terms of their coding potential) in the same cell nucleus are either on or off depending solely on their chromosomal location. One chromosomal site, the *MAT* locus, provides an environment in which genes can be expressed, while other sites, the *HM* loci, provide environments in which they are silenced. Similarly, chromosomal regions adjacent to telomeres provide an environment in which genes that are perfectly well expressed in their normal locations are silenced, although not as efficiently as at the *HM* loci. Accumulating evidence shows that the structure and composition of chromatin is an essential determinant of the silencing environment.

The following properties all indicate that DNA at both telomeres and mating type loci is packaged in a manner that resembles the heterochromatin in higher eukaryotes. First, it is relatively resistant to enzyme modification *in vivo* and to cutting *in vitro* by endonucleases, both of which suggest a modified chromatin structure. (It is interesting to note that the translocation of the mating type genes involves double-stranded cleavage of DNA by an endonuclease. During mating type switching, the *HML* and *HMR* loci are protected from the nuclease until the *MAT* locus, which is not protected, is cleaved; see Fig. 10.14.) Second, it replicates late in S-phase and, finally, it is packaged in nucleosomes that show a relatively low level of acetylation of histones H3 and H4. These properties do not, in themselves, justify the use of telomeres and mating type genes in yeast as models of heterochromatin in general. However, they do improve the chances that results with yeast will be relevant to higher eukaryotes.

Non-histone proteins involved in gene silencing in yeast

A number of genes in *S. cerevisiae*, when mutated, prevent silencing at the mating type loci *HMLα* and *HMRa*. Some of these, for example *RAP1* and *ABF1*, encode proteins known to bind to the E and I silencer elements (Fig. 10.15), while others do not seem to have a direct role in silencing and probably influence mating efficiency through indirect effects. The *NAT* and *ARD* genes provide good examples of this. Loss-of-function mutations in either of these genes cause serious mating defects. However, functional analysis has shown that they encode enzymes that acetylate the N-terminal primary amino group of many newly synthesized proteins. This process is unrelated to the acetylation of specific lysines in the N-terminal domains of the core histones and, as far as we know, plays no role in gene expression. The mating phenotype is likely to be an indirect consequence of their role in cellular protein metabolism. This serves as a useful reminder that the chain of cause-and-effect that connects a mutation with a phenotype may be long and complex. They are not necessarily linked in a way that is functionally relevant or particularly informative. The definitive connection of a particular gene product with a defined cellular function requires analysis and description of the molecular events involved, i.e. genetics must be backed up with biochemistry.

The four genes *SIR1–4* are all necessary for silencing the *HML/HMR* loci and consequently for efficient mating. A combination of genetic and biochemical analysis shows that these proteins are integral components of gene silencing mechanisms in yeast. The protein products Sir2, 3 and 4 are required for silencing at both telomeres and mating type loci. Sir1 appears to be involved in silencing of mating type loci but is not absolutely required (i.e. *SIR1* mutant cells can still mate, just less efficiently than wild type) while it is not necessary for telomeric silencing. None of the Sir proteins binds DNA, but they have been shown to interact with one that does, namely the Rap1 protein that binds to a specific site in E and I silencer elements (Fig. 10.15). This interaction gives the Sir proteins a direct biochemical link with the silencing mechanism. In addition, Sir3 and Sir4 can bind to the N-terminal tail of histone H4 (see below), while Sir2 has been shown to be an NAD-dependent, histone deacetylase.

Histones and silencing

It came as something of a surprise when, some years ago, it was shown that mutants of *S. cerevisiae* with large deletions of the amino-terminal domains of any one of the four core histones were not only viable, but grew in culture just as well as wild type cells. (The mutant histone genes were carried on plasmids using the approach described in Chapter 6.) Mutants carrying deletions of residues 4-14, 4-19, 4-23 and 4-28 in the amino terminal domain of H4 were all viable, although the next largest deletion, 4-34, was not, presumably because it encroaches into the first a-helix of the globular H4 core region and disrupts nucleosome structure.

The two largest deletions do show an increase in doubling time and a lengthening of the G_2 period of the cell cycle. There is also evidence for unfolding of chromatin and/or repositioning of nucleosomes in these mutants. However, the most striking observation was that the mutants all showed reduced mating efficiency, the effect being particularly dramatic in those that were missing all four lysines that can be modified by acetylation. In those H4 mutants in which mating efficiency was severely depressed, the normally silent *HMLα* and *HMRa* genes were both constitutively transcribed and the mutants neither produced nor responded to mating factors. (For reasons discussed later the effect was much stronger at the *HMLα* locus.) The effect was not part of a general disruption of gene regulation.

These observations have been extended by substituting various amino acids for some or all of the amino-terminal lysine residues of H4. Mutants in which all four amino-terminal lysines were substituted by arginine or asparagine were not viable, whereas substitution by glutamine gave a phenotype very similar to the 4-28 deletion mutant (i.e. increased doubling time and severely reduced mating efficiency). Substitution of any of the residues 16–19 of H4 (lys-arg-his-arg) with glycine resulted in de-repression of the *HMLα* and *HMRa* genes. These H4 mutations were found also to have a major effect on telomeric silencing. In contrast, deletions within the H3 amino terminus were able to suppress silencing at telomeres, but had little effect on mating efficiency. These findings are summarized in Fig. 10.16.

How can the very specific effects of deletion or substitution mutants of histones H3 and H4 on gene silencing be explained? One possibility is that the absence or alteration of the amino-terminal region prevents the assembly of the

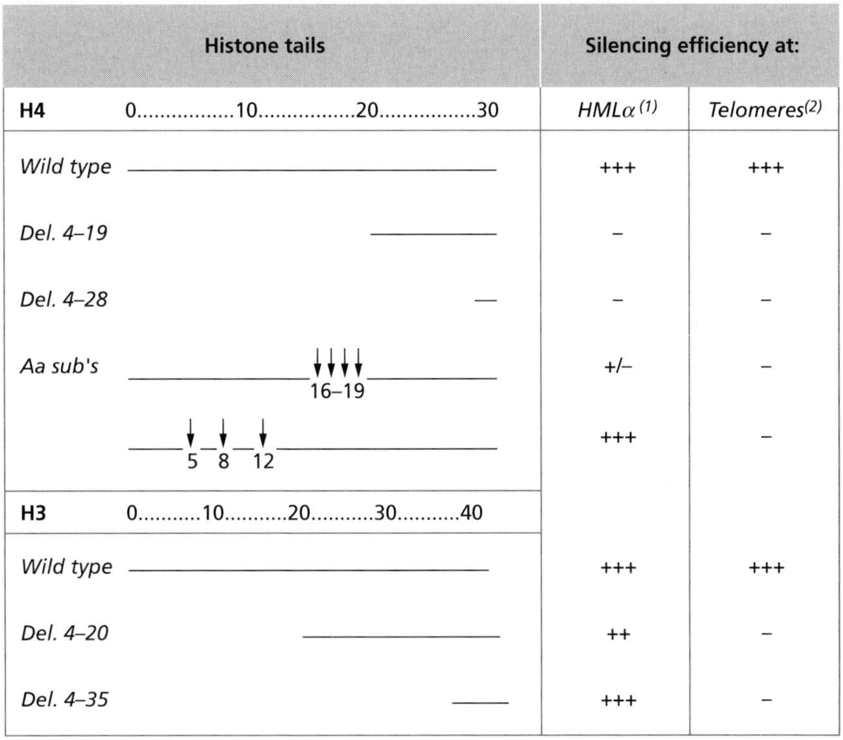

Histone tails		Silencing efficiency at:	
H4 0...........10...........20...........30		*HMLα* [1]	*Telomeres* [2]
Wild type	————————————	+++	+++
Del. 4–19	———————	–	–
Del. 4–28	—	–	–
Aa sub's	↓↓↓↓ 16–19	+/–	–
	↓ ↓ ↓ 5 8 12	+++	–
H3 0.......10.......20.......30.......40			
Wild type	————————————	+++	+++
Del. 4–20	———————	++	–
Del. 4–35	———	+++	–

(1) Assayed by mating efficiency
(2) Assayed by activity of inserted URA3 gene measured by FOA sensitivity

Fig. 10.16 Effects of H3 and H4 mutants on silencing in *S. cerevisiae*.

DNA–protein complex necessary for long-term transcriptional suppression (silencing). A genetic approach to this problem has identified mutations in the *SIR3* gene that are strong suppressors of the H4 16–19 point mutations. That is, cells that carry both an H4 mutation and the *SIR3* suppressor mutation mate normally. This suggests that silencing requires interaction between the Sir3 protein and the amino terminus of H4. Changes in H4 can disrupt this interaction, but parallel changes in Sir3 can restore it (Fig. 10.17). Formally, Sir3 need not bind *directly* to a specific region of the H4 tail, but binding of Sir3 must require a structure (the surface morphology of the nucleosome, for example) that is at least dependent upon the H4 tail. This question has been addressed by *in vitro* experiments testing the ability of recombinant Sir3 and Sir4 proteins to bind to histone H4. Both proteins were shown to bind to H4 (but not other histones) in a way that depended on the presence of the tail domain.

A second factor that can explain the specific effects of the H3/H4 tail mutants is that they are acting to disrupt patterns of histone acetylation that are crucial for efficient gene silencing. This possibility was first suggested by the finding that substitution of acetylatable H3/H4 lysines by neutral residues, such as glutamine, is more effective at disrupting silencing than substitution with arginine, a positive residue that cannot be neutralized *in vivo* by acetylation. This observation suggests that neutralization of specific lysines in the H3/H4 tails, by acetylation for example, can prevent silencing. Consistent with this interpretation, binding of Sir3 and Sir4 proteins to H4 tails *in vitro* appeared to be influenced by acetylation, particularly at lysine 16.

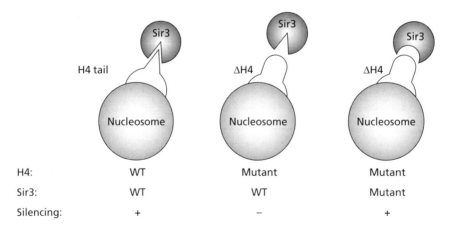

Fig. 10.17 Mutations in Sir3 can restore its ability to bind to nucleosomes with mutations in the N-terminal tail domain of H4. Certain H4 tail mutations can abolish silencing of the mating-type genes in *S. cerevisiae*, leading to decreased mating efficiency. Some mutations in Sir3 restore mating efficiency in these H4 mutants, i.e. they act as suppressor mutations. The altered structure of the mutant Sir3 protein is such that it can now bind to nucleosomes with mutant H4 tails and thereby restore silencing and mating efficiency.

A model for yeast telomeric heterochromatin

A model showing how the major structural components of heterochromatin at yeast telomeres may be assembled is shown in Fig. 10.18. It takes account of experimental data from both genetic and biochemical approaches that demonstrate, or strongly imply, specific protein–DNA and protein–protein interactions. Other proteins may well be involved in the structure shown, and other components are certainly important for heterochromatin assembly and function, but are not integral structural components of the final assembly. The structure is initiated by the binding of Rap1p to the telomeric $C_{1-3}A$ repeats. Rap1 recruits the Sir proteins, which then spread beyond the repeat region by associating with the N-terminal domains of histones H3 and H4, with H3 being particularly important.

Establishment and maintenance of the silent state

In the context of the yeast mating type loci, *establishment* is the process whereby a cell that is expressing HM loci will silence them, while *maintenance* is the stable transfer of the silent state from one cell generation to the next. The two processes differ in their cell cycle dependence. Whereas loss of silencing (i.e. loss of maintenance) can occur at any stage of the cell cycle, re-establishment of the silent state in cells in which it has been lost requires passage through S-phase. This was determined by using temperature-sensitive *SIR3* mutants and switching synchronized mutant cultures from permissive to restrictive temperature at different stages of the cell cycle. It is still not clear what S-phase event is needed for establishment of silencing. It may be DNA replication and ORC binding, although the role of ORC is not limited to S-phase or to the establishment phase.

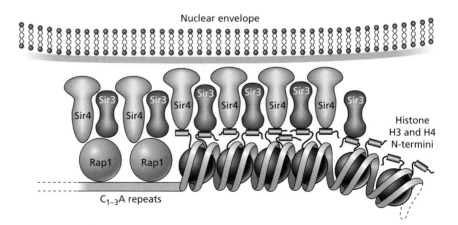

Fig. 10.18 Model for the formation of telomeric heterochromatin at the yeast nuclear periphery. (Redrawn from Grunstein *et al.* 1995, *J. Cell Sci. Supp.*, **19**: 29–36.)

Temperature-sensitive *orc* mutants can cause loss of silencing (i.e. a failure of *maintenance*) in cells arrested in M-phase.

There is evidence to suggest that the Sir1 protein is important for the establishment of silencing but not for its maintenance. In *sir1* mutant cells, the *HM* genes are fully de-repressed in only about 80% of cells; in the remaining 20% they are fully repressed, as in the wild type. This contrasts with *sir2*, *sir3* and *sir4* mutants, in which there is complete loss of *HM* silencing in all cells. With only rare exceptions, the repressed or de-repressed state in *sir1* cells is passed stably from mother to daughter cells. Only once in every 250 or so cell divisions does a de-repressed cell re-establish repression. We can conclude that whereas Sir1 protein is not needed for the *maintenance* of the silent state, it is almost essential for its *establishment*. Significantly, Sir1 protein has been shown to be capable of binding directly to Orc1 protein, the largest subunit of the ORC complex.

The model shown in Fig. 10.19, first put forward by Sternglanz and coworkers, proposes that assembly of the silencing complex at *HMR*-E is initiated by binding of ORC to its cognate DNA binding site. Sir1 links Orc1 with a Sir3–Sir4 protein complex that is further stabilized by association with DNA-bound Rap1. The role of Abf1 is still unclear. Spreading of the silencing complex beyond the silencer element is achieved through association of Sir3/Sir4 with the tail domains of histones H3 and H4, with H4 being particularly important. A similar Sir3/Sir4–H3/H4 interaction is present in the silencing complex at telomeres (Fig. 10.18), although in this case the phenotypes of histone substitution mutants show that it is H3 that has the most influential role.

Spreading

As in the case of PEV discussed in Chapter 9, it is unclear what determines the extent of spreading of the heterochromatic, silencing domain from the telomere along the chromosome. In regions immediately adjacent to the initiation sites,

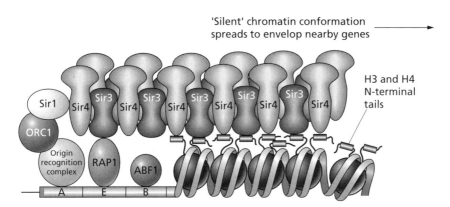

Fig. 10.19 A model for the establishment of silencing at the HMR-E silencer element. (Redrawn from Triolo & Sternglanz 1996.)

spreading is likely to result simply from the binding of Sir proteins, recruited initially by Rap1 or Orc1, to the N-terminal tails of H3 and H4 (Figs 9.18 and 9.19). Spreading could involve a mass action effect regulated by the concentration of one or more of the structural proteins involved. Such a mechanism is consistent with the effect of over-expression of Sir3, which results in the continuous extension of a region of repressive chromatin. Spreading of silencing may also involve the histone deacetylating activity of the Sir2 protein, with a wave of deacetylation extending from the initiation site and facilitating assembly of a heterochromatin-like structure. The fact that the deacetylating activity of Sir2 is NAD dependent (Chapter 8), and therefore potentially influenced by metabolic changes, adds an extra dimension.

It is important to note that telomeres can exert long-range effects that are not explicable by a simple spreading mechanism. The *HMR* locus is normally located 12 kb from the right telomere of chromosome III. If it is moved further away, then it becomes less effectively silenced. If *HMR* is inserted next to an artificial telomere, its silencing ability is enhanced. Thus, the *HMR* locus can interact with an adjacent telomere in a way that enhances its silencing function. (Despite the close similarities between the two loci, *HML* is less sensitive to such telomere position effects.) However, the domain of nuclease-resistant (silenced) chromatin surrounding *HMR* extends only a few hundred base pairs towards the telomere. The intervening chromatin shows normal sensitivity to nucleases, and harbours transcribed genes. If a spreading mechanism is invoked to explain the influence of the telomere on *HMR*, then the chromatin structure that spreads must be one that does not alter nuclease sensitivity and whose effect is exerted only on selected loci. Other mechanisms, such as chromosome looping and/or nuclear positioning, might seem more likely.

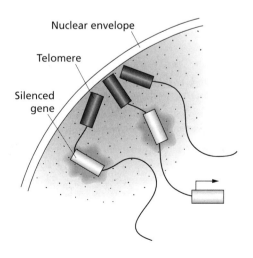

Fig. 10.20 Telomeres in *S. cerevisiae* are located close to the nuclear envelope in regions rich in silencing proteins.

Silencing and position in the nucleus

By using fluorescence *in situ* hybridization (FISH) with DNA probes complementary to telomeric repeats, it can be shown that telomeres are usually clustered adjacent to the nuclear envelope in yeast. This finding is reminiscent of
a long-standing observation that chromosomes in *Drosophila* cells often adopt
the so-called Rabl orientation (named after its discoverer) in which the centromeres are clustered on one side of the nucleus and telomeres at the opposite
side. Immuno-labelling of yeast cells with antibodies to Sir proteins and micro-
scopic analysis has shown that Sir3 and Sir4 are not distributed evenly through
the nucleus, but are present at high concentrations in just those regions where
telomeres are clustered. This is illustrated in Fig. 10.20. A high local concentration of essential protein factors will enhance silencing and provides an attractive
explanation for many experimental results. For example, both *HML* and *HMR*
loci are close to telomeres (12 and 23 kb, respectively) and their silencing efficiency drops if they are moved to more internal chromosome locations. Conversely, in yeast strains carrying plasmids that overproduce Sir proteins, causing
levels to be increased throughout the nucleus, then silencing becomes less
dependent on proximity to telomeres. All of these observations can be understood on the basis of the model illustrated in Fig. 10.20. It seems that assembly
of the structure illustrated in Fig. 10.18 is somehow necessary either for targeting telomeres to the nuclear envelope or for keeping them there. Clustering of
telomeres at the nuclear periphery is lost in strains carrying deletions of the N-
terminal tails of H3 (Δ4-20) or H4 (Δ4-28) and in which linking of chromatin to
the Sir3/Sir4 complex will not occur. Deletions of the H2A or H2B tails are without effect. Note also that high local concentrations of Sir proteins will not necessarily silence all genes in the vicinity. Only those genes that are enveloped by the
spreading Sir complex will be silenced. It is quite possible for active genes to lie
between the telomere and the *HMR* locus.

Proximity to the nuclear envelope can, in itself, enhance silencing. This has
been demonstrated by constructing yeast strains in which DNA constructs containing *HMR* silencer elements were be artificially tethered to the nuclear envelope. Several mutated *HMR* silencers were identified that showed little or no
ability to silence an adjacent reporter gene, *unless* they were tethered to the
nuclear envelope. Silencing was dependent on the Sir proteins, i.e. did not
occur in strains in which one or other of these proteins was absent or nonfunctional through mutation. The results are consistent with the model in Fig. 10.20
but demonstrate that the primary role of the telomere is probably to target the
HMR locus to the nuclear envelope. It may exert no other effect.

Further Reading

DNA methylation and CpG islands

Bird, A. (1986) CpG-rich islands and the function of DNA methylation. *Nature,* **321**: 209–213.
Bird, A. (1999) DNA methylation *de novo. Science,* **286**: 2287–2288.
Craig, G. M. & Bickmore, W. A. (1994) The distribution of CpG islands in mammalian chromosomes. *Nature (Genetics),* **7**: 376–382.
Laird, P. W. & Jaenisch, R. (1996) The role of DNA methylation in cancer genetics and epigenetics. *Ann. Rev. Genet.,* **30**: 441–464.
Tazi, J. & Bird, A. (1990) Alternative chromatin structure at CpG islands. *Cell,* **60**: 909–920.
Yoder, J. A., Walsh, C. P. & Bestor, T. H. (1997) Cytosine methylation and the ecology of intragenomic parasites. *Trends Genet.,* **13**: 335–340.

DNA methylation and transcription

Bird, A. P. & Wolffe, A. P. (1999) Methylation-induced repression-belts, braces and chromatin. *Cell,* **99**: 451–454.
Pikaart, M. J., Recillas-Targa, F. & Felesenfeld, G. (1998) Loss of transcriptional activity of a transgene is accompanied by DNA methylation and histone deacetylation and is prevented by insulators. *Genes Dev.,* **12**: 2852–2862.
Selker, E. (1999) Gene silencing: repeats that count. *Cell,* **97**: 157–160.

Methyl-DNA-binding proteins

Chen, R. Z., Akbarian, S., Tudor, M. & Jaenisch, R. (2001) Definiciency of methyl-CpG binding protein 2 in CNS neurons results in a Rett-like phenotype in mice. *Nature Genetics,* **27**: 327–331.
Guy, J., Hendrich, B., Holmes, M., Martin, J. E. & Bird, A. (2001) A mouse *Mecp2*-null mutation causes neurological symptoms that mimic Rett syndrome. *Nature Genetics,* **27**: 322–326.
Hendrich, B. & Bird, A. (1998) Identification and characterization of a family of mammalian methyl-CpG binding proteins. *Mol. Cell Biol.,* **18**: 6538–6547.
Nan, X., Campoy, J. & Bird, A. (1997) MeCP2 is a transcriptional repressor with abundant binding sites in genomic chromatin. *Cell,* **88**: 471–481.

The enzymology of DNA methylation

Lei, H., Oh, S. P., Okano, M. *et al.* (1996) *De novo* DNA cytosine methyltransferase activities in mouse embryonic stem cells. *Development,* **122**: 3195–3205.
Okano, M., Xie, S. & Li, E. (1998) Cloning and characterization of a family of novel mammalian DNA (cytosine-5) methyltransferases. *Nature (Genetics),* **19**: 219–220.
Wolffe, A. P., Jones, P. L. & Wade, P. A. (1999) DNA demethylation. *Proc. Natl. Acad. Sci., USA,* **96**: 5894–5896.

DNA methylation and chromatin modification

Cameron, E., Bachman, K. E., Myöhänen, S., Herman, J. & Baylin, S. B. (1999) Synergy of demethylation and histone deacetylase inhibition in the re-expression of genes silenced in cancer. *Nature (Genetics),* **21**: 103–107.
Ng, H.-H. & Bird, A. (1999) DNA methylation and chromatin modification. *Curr. Opin. Genet. Dev.,* **9**: 158–163.

Silencing in yeast

Laurenson, P. & Rine, J. (1992) Silencers, silencing and heritable transcriptional states. *Microbiol. Rev.*, **56:** 543–560.
Loo, S. & Rine, J. (1999) Silencing and heritable domains of gene expression. *Ann. Rev. Cell Dev. Biol.*, **11:** 519–548.
Zakian, V. A. (1989) Structure and function of telomeres. *Ann. Rev. Genet.* **23:** 579–604.

Histones and silencing

Hecht, A., Laroche, T., Strahl-Bolsinger, S., Gasser, S. & Grunstein, M. (1995) Histone H3 and H4 N-termini interact with SIR3 and SIR4 proteins: a molecular model for the formation of heterochromatin in yeast. *Cell*, **80:** 583–592.
Thompson, J. S., Hecht, A. & Grunstein, M. (1993) Histones and the regulation of heterochromatin in yeast. *Cold Spring Harbor Symp. Quant. Biol.*, **43:** 247–256.
Thompson, J. S., Ling, X. & Grunstein, M. (1994) Histone H3 amino terminus is required for telomeric and silent mating locus repression in yeast. *Nature*, **369:** 245–247.
Wright, J. H., Gottschling, D. E. & Zakian, V. A. (1992) *Saccharomyces* telomeres assume a non-nucleosomal chromatin structure. *Genes Dev.*, **6:** 197–210.

Non-histone proteins and silencing

Andrulis, E. D., Nelman, A. M., Zappulla, D. C. & Sternglanz, R. (1998) Perinuclear localization of chromatin facilitates transcriptional silencing. *Nature*, **394:** 592–595.
Cockell, M., Palladino, F., Laroche, T. *et al.* (1995) The carboxy termini of Sir4 and Rap1 affect Sir3 localization: evidence for a multicomponent complex required for yeast telomeric silencing. *J. Cell Biol.*, **129:** 909–924.
Moretti, P., Freeman, K., Coodly, L. & Shore, D. (1994) Evidence that a complex of SIR proteins interacts with the silencer and telomere-binding protein RAP1. *Genes Dev.*, **8:** 2257–2269.
Strahl-Bolsinger, S., Hecht, A., Luo, K. & Grunstein, M. (1997) SIR2 and SIR4 interactions differ in core and extended telomeric heterochromatin in yeast. *Genes Dev.*, **11:** 83–93.
Triolo, T. & Sternglanz, R. (1996) Role of interactions between the origin recognition complex and SIR1 in transcriptional silencing. *Nature*, **381:** 251–253.

Chapter 11: Cellular Memory and Imprinting

Introduction

Cells are sometimes said to have the ability to remember who and what they are. To anyone familiar with the growth of cells in tissue culture, this would seem to be self-evident. Cultured skin fibroblasts, unless treated in special ways, remain fibroblasts through successive passages until the culture eventually senesces and dies. They do this even though they have all the genes needed to become perfectly good neuroblasts, hepatocytes or any other cell type. The same can be said of many other cultured cell types, including transformed cell lines, many of which have been grown in laboratories around the world for years, without major changes in their morphological and functional characteristics. We can conclude from this that cells, unless told to do otherwise, tend to stay as they are.

As the differences between cell types result from their different patterns of gene expression, 'staying as you are', for a cell, means remembering a particular pattern of gene expression from one cell generation to the next. Specifically, the cell must have a mechanism by which individual genes retain the same level of expression – in the simplest case either on or off – through successive cell cycles. This has been called cellular memory and is not a trivial problem. Active and inactive genes must be marked in a way that will survive both the passage of the DNA replication complex during S-phase and the drastic chromatin restructuring that accompanies passage through mitosis. For some genes, the so-called 'housekeeping' genes that are essential for the survival and growth of all cell types, there is no decision to be made: they are always transcribed. For these genes, the information that determines their consistent expression may lie in the DNA sequence itself, although this is not known for certain. However, for tissue-specific genes (which constitute the great majority of genes) the same DNA sequence is expressed in some cell types but not in others. For this simple reason, patterns of gene expression must be inherited from one cell generation to the next by mechanisms that lie outside the DNA sequence itself. This is referred to as *epigenetic* inheritance.

For a small number of genes, epigenetic mechanisms can determine expression not only from one cell generation to the next, but also from one generation of the organism to the next, i.e. through the germ line. These genes are said to be *imprinted* and their expression, or lack of it, is dependent on whether they were inherited paternally (i.e. from the father through the sperm) or maternally (i.e. from the mother through the ovum). For example, in many tissues of the developing foetus, the paternal copy of *Igf2*, a gene that encodes a foetal growth factor, is transcribed while the maternal copy is not. Once again, the difference cannot lie in the DNA sequence. Imprints are erased and reset every time the gene passes through the male or female germ line (see below), so an allele that is paternal (and expressed) in one generation may be maternal (and silent) in the next. Epigenetic mechanisms must be involved.

It could reasonably be argued that the unchanging maintenance of cell-specific patterns of gene expression seen in cultured cells is not typical of the situation *in vivo*. In the body (the 'real' world for cells), generation of cells of a specific type is often achieved not by growth and division of existing cells of that type, but by inducing stem cells to begin differentiating down the appropriate pathway. The ongoing replenishment of the epithelial cells lining the intestine by differentiation and migration of stem cells located deep in the tissue provides a good example of this. Only when cells are removed from the signals that enable them to progress along the differentiation pathway (by placing them in tissue culture for example) or when they fail to respond to such signals (through mutation), do they grow and divide without the accompanying differentiation. However, there is still a need for a genetic memory mechanism. Each step along the differentiation pathway builds on the preceding one. With each step along the pathway, some genes will change expression, but many will not. This is illustrated by the hypothetical example in Fig. 11.1. Three genes, numbered 1–3, are shown and their expression at each stage in the differentiation pathway indicated as + or –. At stage II, gene 1 is switched on and gene 2 is switched off, and these patterns of expression are maintained through stages III and IV. In contrast, gene 3 is switched on at stage II, maintained in this state until stage IV and then switched off again. Differentiation requires mechanisms for switching gene expression at the appropriate stage and for maintaining (remembering) these transcriptional states through several cell generations until another switching signal is received. The two mechanisms may be quite different.

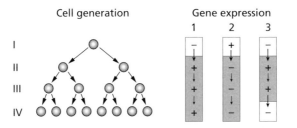

Fig. 11.1 Switching and maintenance of gene expression are both necessary for differentiation. The three genes shown all switch their transcriptional state from generation I to generation II (heavy arrows). Genes 1 and 3 switch from *off* to *on* and 2 switches from *on* to *off*. Genes 1 and 2 maintain their respective states through succeeding generations (shading). Gene 3 maintians its *on* state for only two generations and is then switched *off*.

Maintenance of transcriptional states

Particularly good examples of the biological role of epigenetic memory are provided by the highly conserved gene families whose job is to ensure that specific tissues and cell types appear in the right place and at the right time during embryonic development. These genes have been most extensively studied in *Drosophila*, but close homologues are found in other organisms, including vertebrates. Among its many advantages as an experimental organism, the fruitfly has a body that is built up of visibly defined segments (Fig. 11.2). Specific segments are associated with particular structures, i.e. legs, wings and such. This segmental pattern is laid down very early in development and the trained eye

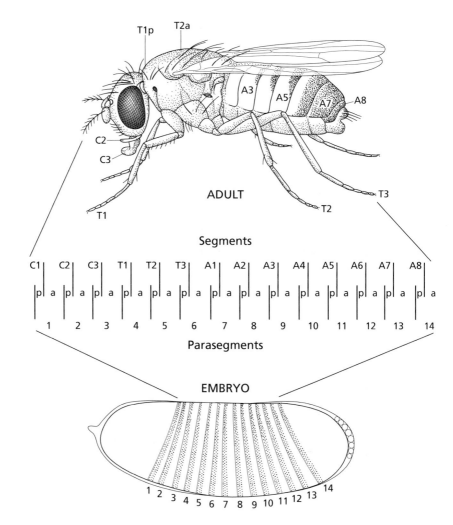

Fig. 11.2 Segments and parasegments in the fruitfly *Drosophila*. (Redrawn from Lawrence, 1992.)

can detect a series of grooves circling the stage 10 embryo (5–6 h after fertilization). The regions delineated by these grooves are known as *parasegments*. The reason for this designation is that the parasegments are not exactly in register with the body segments of the adult fly. In fact, each body segment develops from the anterior part of one parasegment and the posterior part of the adjacent parasegment (Fig. 11.2). This complication need not bother us and in what follows we will think in terms of the parasegments.

For the purposes of the present discussion, the most interesting property of the parasegments is that they differ in their patterns of gene expression. Indeed, the easiest way to distinguish the different parasegments is to label the embryo with antibodies to proteins that are made only in particular parasegments or subsets thereof. An example is shown in Fig. 11.3. Using this approach, parasegments can be detected even earlier in development, well before there is any morphological evidence of their existence. It is, of course, these different patterns of gene expression that provide the molecular basis for the different developmental pathways followed by the different parasegments. Two families of genes, known as the 'gap' and 'pair rule' genes, are particularly important in these early stages. Their function, in general terms, is to regulate the expression of other genes. These include both other members of the gap and pair rule families and, crucially, the *homeotic* genes, the master regulators of development.

The homeotic genes illustrate the importance of stabilizing patterns of gene expression through development

The homeotic genes are defined as genes whose mutation results in transformations of one body part into another. Sometimes these transformations are spectacular. For example, the *Antennapaedia* gene is normally expressed in

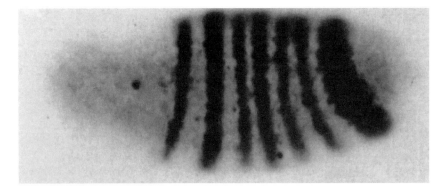

Fig. 11.3 An autoradiograph showing the location of mRNA from the pair-rule gene *fushi tarazu* (*ftz*) in a *Drosophila* embryo. The *ftz* gene is expressed in each of the even-numbered parasegments 2 to 14 (see Fig. 11.2). (Reproduced from MacDonald & Struhl 1986, *Nature,* **324**: 537–545.)

parasegments 3–6, which develop into the thoracic regions of the adult fly (Fig. 11.2). Inappropriate expression of *Antennapaedia* in a more anterior parasegment can cause this region to develop as if it were a thoracic segment, resulting, famously, in the appearance of a leg where an antenna should be. Such 'homeotic' transformations, identified by geneticists long before the molecular properties of these genes or their products were unravelled, support the concept of a family of master regulator genes whose products determine the developmental pathway to be followed by defined groups of cells.

There are two clusters of homeotic, master regulator genes in *Drosophila*, namely the *Bithorax* Complex (BX-C, with three genes) and *Antennapaedia* Complex (ANT-C with five). Genes of the BX-C are expressed in parasegments 5 to 14 and the ANT-C genes in parasegments 1 to 6. Each individual gene has its own expression pattern. Expression of the three BX-C genes is shown in Fig. 11.4. *Ubx* is expressed in parasegments 5 to 13, *abdA* in parasegments 7 to 13 and *AbdB* in parasegments 10 to 14. Misexpression of these genes transforms the developmental fate of the parasegments. For example, if a mutation causes *Ubx* expression to extend into parasegment 4, then this parasegment follows the developmental pathway appropriate to parasegment 5, where *Ubx* is normally expressed (Fig. 11.4). The most obvious manifestation of this is the absence of the wing and its replacement with a haltere, a small projection with a probable role in flight control (Fig. 11.2). The protein products of the homeotic genes contain a particular DNA-binding motif, the homeobox (see Chapter 1). The sequence-specific DNA-binding properties conferred by the homeobox enable the proteins that have it to act as transcription factors, controlling the

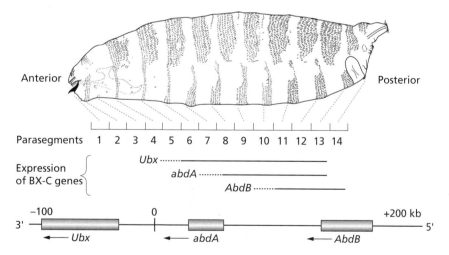

Fig. 11.4 Expression of genes of the Bithorax Complex in parasegments of the *Drosophila* embryo. The bithorax complex (BX-C) contains three genes, *abdominal A* (*abdA*), *abdominal B* (*AbdB*) and *Ultrabithorax* (*Ubx*). Their arrangement in the complex and directions of transcription are shown in the lower part of the figure, and the parasegments in which they are expressed in the central part.

expression of other groups of genes. In the case of the homeotic gene products, the genes that are regulated are those defining developmental pathways. (Note that the homeobox is not confined to the protein products of the homeotic genes; there are over 100 proteins in *Drosophila* that contain a homeobox.)

For both the ANT-C and BX-C genes, the changes in expression as one moves from the posterior to the anterior end of the embryo correspond to the order of the genes along the chromosome (Fig. 11.4). Also, once switched off, a gene is not re-expressed in a more anterior segment. This has given rise to the interesting idea that a change in chromatin structure might spread along the chromosome, progressively enveloping, and silencing, each of the genes in the two complexes. Thus, in parasegment 9 (and all the more anterior parts of the embryo), *AbdB* will be packaged in 'silencing' chromatin, in parasegment 7 the structure will have spread to *abdA* and in parasegment 4, *Ubx* will have been included. However, the idea that 'silencing' chromatin spreads from a single point 5′ of *AbdB* to envelop all three genes is not supported by the finding that translocating the *Ubx* gene to a separate chromosome produces normal flies. The gene can clearly function normally outside its usual genomic location. This finding requires that each of the three genes is associated with motifs that can, if necessary, initiate silencing independently and at the appropriate time (developmentally) and place. Nonetheless, it is a striking fact that genes in mice and other species that are homologous to members of the ANT-C and the BX-C genes in *Drosophila* generally show the same relationship between their order on the chromosome and their pattern of expression in the developing embryo. This is sometimes referred to as the *co-linearity* phenomenon.

Regulation of the BX-C genes

In view of the dire consequences of mis-expression of the ANT-C and BX-C genes, it is not surprising that their regulation is complex and subject to numerous checks and balances. We will focus here on BX-C, but the general principles that emerge are applicable to the ANT-C genes as well.

Extensive mutational analysis has revealed DNA regions across the BX-C that are important in the regulation of each of the three genes. These DNA elements are shown in Fig. 11.5. It is clear that, of the 300 kb of the BX-C, far more is taken up by regulatory sequences than by coding DNA. Regulatory DNA elements can be located upstream or downstream of the gene they regulate, or within an intron of that gene (Fig. 11.5). Each regulatory element is used in a particular parasegment. For example, iab5, iab6, iab7 and iab8–9 are used to determine the expression of *AbdB* in parasegments (PS) 10, 11, 12 and 13–14, respectively. Expression is weak in PS10 and strongest in PS13–14. Once again, the order of the regulatory elements along the chromosome corresponds to the order of the parasegments in which they are used.

The regulation of *AbdB* in parasegments 10, 11 and 12 is shown diagrammatically in Fig. 11.6. Regulatory elements are 'activated' by the actions of

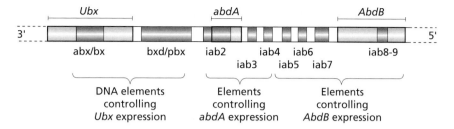

Fig. 11.5 DNA elements control expression of the component genes of the bithorax complex. Individual genes are shown as lightly-shaded boxes and control elements as darkly-shaded boxes. Control elements can occur upstream or downstream of the genes they regulate, or within the gene. Control elements have been defined genetically through the properties of their mutants.

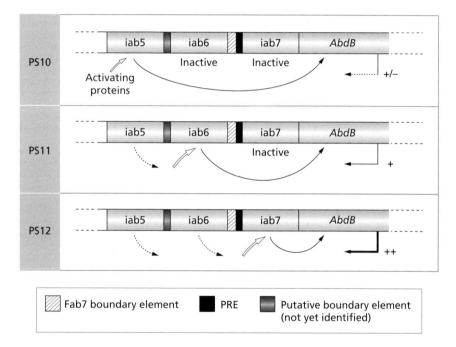

Fig. 11.6 Regulation of *AbdB* expression in parasegments 10, 11 and 12 in wild type flies. The long, curved arrows indicate that the control element exerts an influence on expression of the *AbdB* gene. The mechanism by which this is achieved remains unclear. The hatched box is the *Fab7* boundary element and the filled box is a polycomb response element (PRE). It is likely that a boundary element separates the iab5 and iab6 control elements (shaded box), but it has not yet been identified.

parasegment-specific combinations of proteins, mainly products of the gap and pair-rule genes. Once activated, the regulatory element imposes a pattern of expression on *AbdB* that is appropriate to the parasegment. This requires interaction over large distances. For example, iab6 is separated from the *AbdB* promoter by about 80 kb of DNA. Interestingly, the strongest expression is induced by the most promoter proximal elements. We will return later to possible mechanisms for this regulation. If one of the regulatory elements is deleted, then control of *AbdB* expression is taken over by its 3′ neighbour. Thus, when iab7 is deleted, regulation of *AbdB* in PS12 is taken over by iab6, and PS12 therefore takes on the developmental properties of PS11 (Fig. 11.7). This tells us that iab6 can be activated, early in development, even in regions of the embryo that, under normal circumstances, would develop into PS12. In view of this, it is likely that, in normal embryos, iab6 remains in an active (or potentially active) state in PS12, but that its activating effect is overwhelmed by the much stronger effect of iab7.

Genes of the polycomb and trithorax groups encode essential regulatory proteins

The control of homeotic genes such as *AbdB* has two quite separate component parts, namely initiation and maintenance. The reason for this split is that the gap and pair-rule genes, whose products are responsible for initiating activity, are expressed only transiently early in development. In contrast, the homeotic genes are expressed (strictly where required) throughout development. To allow this to happen, mechanisms must exist that lock in place the expression patterns that are set early in development. Stabilization of the transcriptional status of homeotic genes through development is provided by the protein products of two gene families, namely the polycomb group (PcG) and the trithorax

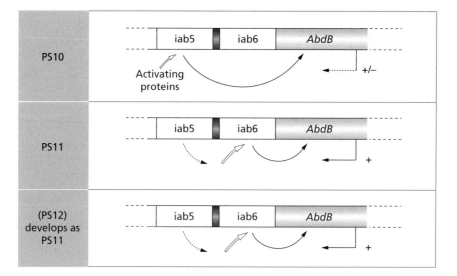

Fig. 11.7 Regulation of *AbdB* expression in parasegments 10, 11 and 12 in mutant flies lacking the iab7 control element.

group (trxG). Genes have been assigned to either family largely on the basis of their genetic properties, i.e. the phenotypes of their mutants, in *Drosophila*. Such experiments gave rise to the idea that PcG proteins maintain states of transcriptional inactivity (silencing) whereas trxG proteins maintain chromatin states that are permissive for transcription. For example, in flies carrying mutations of PcG genes, expression of homeotic genes early in development is normal, but inappropriate expression of genes that should be silent occurs later, i.e. *initiation* is normal but *maintenance* is not.

All PcG genes, when mutated, cause developmental defects that are attributed to inappropriate expression (specifically, lack of silencing) of homeotic genes, and several PcG genes are named on the basis of the phenotypic effects they produce when mutated (e.g. *posterior sex combs, Psc*). There are now more than a dozen members of the PcG gene family, some of which are listed in Table 11.1. The protein products of PcG genes exhibit, collectively, a variety of structural motifs such as chromodomains, zinc fingers and RING fingers, all of which can be involved in protein–protein interactions. There is substantial evidence that PcG proteins exert their effects by assembling as large, multiprotein complexes. For example, different members of the family co-localize by immunofluorescence microscopy and co-immunoprecipitate. Surprisingly, in view of their function, the PcG proteins, with only one exception (*pleiohomeotic, pho*), do not have the ability to bind DNA. There is little or no sequence homology between different PcG genes, but most of them have close homologues in mammals (Table 11.1).

Table 11.1 Examples of polycomb group (PcG) and trithorax group (TrxG) genes and proteins.

Drosophila	Mammals	Protein motifs	Protein size (kDa)	Properties
PcG genes				
Pc	*M33* (human)	Chromodomain	44	
Psc	*Bmi-1, Mel-18*	RING finger	170	Mutations in mammals cause homeotic transformations
E(z)	*EZH1, EZH2* (human)	SET domain	87	
Su(z)2	*Bmi-1, Mel-18*	RING finger	144	Mammalian proteins bind specific DNA sequences
Esc	*Eed*	WD40 repeats	48	Mutations in mammals alter anterior–posterior patterning
TrxG genes				
Trx	*MLL, mll*	SET domain, PHD fingers		DNA methyltransferase domain in mammals
Brm	*BRG1, brg1*	Bromodomain		DNA helicase, homology to yeast SWI2/SNF2 chromatin remodelling proteins
Trl				GAGA factor, DNA binding

The trxG genes are defined by the finding that, when mutated, they cause homeotic transformations that are attributable to *loss* of expression of homeotic master regulator genes such as *Ubx* (i.e. to a failure to maintain the active state). TrxG mutations tend to counteract the effects of PcG mutations. Like the PcG proteins, the trxG proteins are heterogeneous, but most (perhaps all) have close homologues in mammals. Some examples are listed in Table 11.1.

PcG response elements

A variety of genetic and biochemical approaches has been used to try and define the DNA sequences with which complexes of PcG and trxG proteins are functionally associated. Most information so far available concerns the PcG proteins and we will focus on these. DNA sequences that can assemble functional PcG complexes are known as PcG response elements (PREs). They can be up to several hundred base pairs in length. A PRE, when included in a DNA construct containing a reporter gene, will lead to silencing of that gene in cells in which PcG proteins are available. As expected, PREs are distributed across the BX-C and ANT-C, but immunostaining polytene chromosomes with antibodies to the PcG protein PC has revealed approximately 100 binding sites scattered across the *Drosophila* genome. PcG proteins must have roles beyond the control of homeotic genes. The same immunolabelling approach has shown that trxG proteins are also widely distributed across polytene chromosomes. PcG and trxG proteins sometimes co-localize, although this does not mean that they necessarily bind to the same DNA sequences.

There are no large-scale similarities in DNA sequence between different PREs, but short conserved sequence motifs have been identified. Significantly, the DNA sequence motif recognized by the *pho* protein is found in many PREs, suggesting that *pho* binding to DNA may be central to the targeting and assembly of many PcG complexes. However, PREs in general are likely to be modular in their binding properties, containing motifs for a variety of DNA-binding proteins that anchor PcG proteins.

Spreading of PcG-mediated silencing

PcG-induced silencing is known to cover large genomic distances, up to 300 kb in the case of the bithorax complex. This raises the question of how silencing spreads from the PRE. The most obvious possibility is that a protein complex, including PcG proteins and others, simply spreads from the PRE to coat the chromosome. Similar models have been put forward to explain heterochromatin spreading in position effect variegation (Chapter 9). In order to address this question, chromatin from *Drosophila* cells was cross-linked with formaldehyde to bind the associated proteins tightly to DNA, broken into small fragments by sonication and immuno-precipitated with antibodies to the PC protein. (The technique is described in Box 2, p. 112). DNA sequences in the antibody-bound fraction must have been closely associated with PcG proteins, and by identifying these sequences, the extent of PcG spreading across specific genomic regions can be determined. This approach was used to study the distribution of PcG proteins across the bithorax complex in a cultured cell line derived from *Drosophila* embryos. PcG proteins were found associated with DNA tens of kilobases beyond the PRE itself. This and the fact that PcG proteins form large, multi-subunit complexes in *Drosophila* and other species, are both consistent with a spreading mechanism.

It is worth recalling here that, as discussed in Chapters 8 and 9, silencing can be achieved in different ways. It is possible that, rather than employing spreading (or looping) mechanisms, PcG binding helps locate specific regions of the genome in subnuclear compartments where transcription is not favoured, possibly due to exclusion of crucial components of the transcription machinery. (Figures 9.11, 9.12 and 10.20 illustrate situations where intranuclear positioning seems to play a crucial role in gene silencing.) If diploid, interphase cells are immunolabelled with antibodies to PcG or trxG proteins, the proteins are seen to be located as discrete patches in the nucleus. Some patches contain both sets of proteins. This clustering has led to the suggestion that PcG and trxG proteins may be involved in the large-scale organization of chromatin in the nucleus, repositioning specific genes or control elements into nuclear locations where their active or inactive state can be maintained. This idea is lent support by the finding that mutations to DNA elements that have a 'boundary' function, preventing the spread of active or inactive states from one domain to the next (see below), also disrupt the intranuclear location of PcG proteins.

Fab7: a DNA element that blocks spreading of transcriptional states

While it is necessary for PcG and trxG proteins to act over large distances, it is equally important that their effects are limited to defined chromatin domains. Functionally, their effects must not spread randomly across parasegment boundaries. It seems that, in addition to the regulatory elements discussed so far, there are others whose role is to prevent this from happening. A 4.3 kb region designated Fab7 and located between iab6 and iab7 in the BX-C (Figs 11.5, 11.6) provides a good example. In mutant flies in which Fab7 has been deleted, both iab6 and iab7 become active in PS11, rather than just iab6. The result is that iab7 determines the expression pattern of *AbdB* and PS11 develops in the same way as its (posterior) neighbour PS12, i.e. there is a homeotic transformation (Fig. 11.8). This result can be explained by proposing that, in the absence of Fab7, the active state of iab6 has spread into iab7. If the deletion of Fab7 includes the iab7 PRE located close to it, then *all* cells in PS11 express *AbdB* under the control of iab7. However, if the deletion leaves the PRE in place, then in a small subpopulation of PS11 cells, *AbdB* is not expressed detectably, presumably because iab6 and iab7 are *both inactive*. The likely explanation for this is that in PS11, in the absence of Fab7, there is a competition between the active

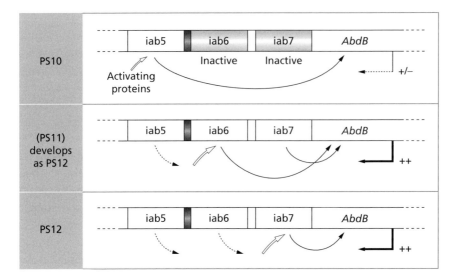

Fig. 11.8 Regulation of *AbdB* expression in parasegments 10, 11 and 12 in mutant flies lacking the *Fab7* boundary element.

state, attempting to spread from iab6, and the inactive state, attempting to spread from iab7. In the absence of the iab7 PRE, the active state always wins (as shown in Fig. 11.8). When the iab7 PRE is intact, it strengthens the ability of iab7 to form the inactive state by allowing assembly of PcG proteins, with the result that *in some cells* both domains are *inactive*. This result is important because it shows that Fab7 can act as a boundary for both the active and inactive state.

These results are consistent with Fab7 having a role in preventing cross-talk between regulatory elements, i.e. acting as a boundary element. However, if this is so, then we are entitled to ask why it is that Fab7 does not prevent interaction between regulatory elements and the *AbdB* gene itself. Iab5 in PS10 and iab6 in PS11 (Fig. 11.6) are both able to activate *AbdB* across the putative Fab7 boundary. It may be that the mechanisms by which iab5–7 interact with *AbdB* are quite different from those responsible for spreading of active or inactive states between regulatory domains. In fact, it is important to remember that, although the convenient term 'spreading' has been used throughout this discussion, it should not be *assumed* that a protein complex literally spreads along the chromosome from the PRE (though this may happen). The battle between iab6 and iab7 in mutants lacking the Fab7 element but retaining the iab7 PRE may reflect competing attempts to drag the two domains to different nuclear locations.

The chromatin packaging of the Fab7 element is informative. It contains both positioned nucleosomes and a set of DNaseI hypersensitive sites (Fig. 11.9). These are usually associated with DNA regions involved in protein binding (see Chapter 6). The presence of nucleosomes immediately distinguish-

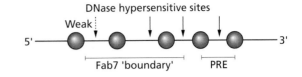

Fig. 11.9 Chromatin structure of the Fab7 element and the iab7 polycomb response element (PRE).

es Fab7 from the first boundary elements to be identified, namely the *scs* and *scs'* regions that mark the limits of the *hsp70* heat shock locus. Although nuclease resistant, these regions are not packaged as canonical nucleosomes, hence their designation as regions of special chromatin structure, *scs*.

In view of the importance of the Fab7 boundary, it would be reasonable to conclude that there must be boundary elements between other regulatory elements. As yet, only one other has been identified (between iab4, a regulator of *abdA*, and iab5). While there must be some mechanism for preventing crosstalk between the regulatory elements, it remains to be seen whether this always operates in the same way as Fab7.

The molecular dissection of PcG-silencing mechanisms

In an attempt to define more closely the molecular basis of polycomb silencing, an artificial construct was made in which a 3.6 kb sequence incorporating Fab7 and the adjacent PRE was attached to a GAL4 dependent promoter, a *lacZ* reporter gene and a mini-*white* gene. The construct is shown in Fig. 11.10. A series of *Drosophila* lines was established in which this construct was inserted stably into the fly genome. Because the strains used for transformation lacked a functional *white* gene, they had white eyes, rather than the normal red. (Remember that in *Drosophila*, genes are named after the *mutant* phenotype; thus, the 'white' gene, when working *normally*, gives red eyes.) So, the flies that had the construct would have red eyes only if the inserted mini-*white* gene was working. Transcription of the mini-*white* and *lacZ* genes requires the GAL4 activator. This bacterial transcription factor is not normally present in flies, but can be provided, for the purposes of this and related experiments, by a second construct engineered into the genome of specific fly strains. Production of GAL4 can be regulated accurately by using various promoters and developmentally controlled enhancers. Fly strains are available that can generate GAL4 in appropriate amounts through development. When the test construct, lacking the Fab7 element, was stably inserted into these strains, not only was β-galactosidase produced in all tissues, but the adult flies had either completely or partially red eyes. Thus, in GAL4-producing strains, the construct can produce both β-galactosidase, from *lacZ*, and normal eye colour, via the mini-*white*, albeit not as efficiently as in wild type flies. When the Fab7, PRE-containing element was included in the construct, both genes were silenced, i.e. β-

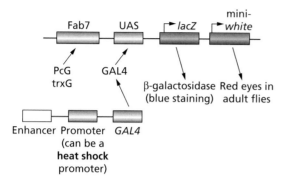

Fig. 11.10 DNA constructs that can be used to test the role of Polycomb Group (PcG) and trithorax Group (trxG) proteins in regulating expression of *lacZ* and *white* genes under the control of a GAL4 promoter.

galactosidase levels were low and flies were white or yellow eyed. This confirms that Fab7 plus the PRE can act as a silencing element.

The Fab7 construct can be switched from a repressed to a stable, active state

Fab7/PRE-mediated silencing depends on competition between GAL4 and PcG proteins. This was shown by testing fly strains in which the *GAL4* gene was under the control of a heat-shock promoter (Fig. 11.10). If the amount of GAL4 was massively increased by heat shock, then silencing was lost.

Fly strains in which GAL4 was inducible by heat shock allowed extension of the Fab7 experiments to investigate developmental regulation of the mechanisms that maintain the silent or active states. If embryos from such strains were subjected to a brief heat shock (30 min at 37°C) and then allowed to develop at their normal growth temperature (18°C), adult flies showed an increased frequency of red eye colour (Fig. 11.11). For this to happen, the relief of inhibition induced by the transient increase in GAL4 must have been maintained through many cell generations, long after GAL4 levels had fallen away. This effect was seen only if the heat shock was applied during the early embryonic stage of development. Brief heat shock during the larval stages could transiently induce the *lacZ* gene, but did not lead to *stable* de-repression of either *lacZ* or the mini-*white* gene (Fig. 11.11). Although the proteins involved in the stabilization of the transcriptional states of *lacZ* and mini-*white* are still not known, the system lends itself to experiments to identify them. For example, maintenance of the active state seems less efficient in fly strains with mutations in trxG protein GAGA factor, suggesting (perhaps not surprisingly) that this protein is involved.

Germ line inheritance of an epigenetic state

Even though this system is an artificial one, the experiments make an important general point. An environmental change during early embryogenesis (i.e. a temperature increase with, in this experiment, induction of a transcription

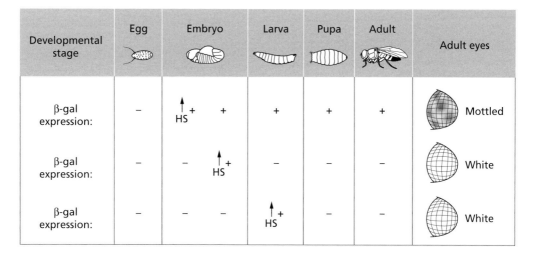

Developmental stage	Egg	Embryo	Larva	Pupa	Adult	Adult eyes
β-gal expression:	–	↑+ HS +	+	+	+	Mottled
β-gal expression:	–	–	↑+ HS	–	–	–	White
β-gal expression:	–	–	–	↑+ HS	–	–	White

Fig. 11.11 Expression of *lacZ* (β-galactosidase) and *white* (eye colour) genes from the Fab7 construct (Fig. 11.10), following induction of GAL4 (by heat shock, HS) at different stages of development. If a transient burst of GAL4 is provided early in development, β-galactosidase is detected throughout development, and the *white* gene is expressed in adult eyes. A burst of GAL4 later in development causes only transient expression of β-galactosidase and no expression of *white*.

factor) can alter the transcriptional potential of a gene in a way that is inherited stably through many cell generations, even in the absence of the original inductive signal. But the really surprising result was the observation that the de-repression induced by the burst of GAL4 could be inherited through *meiosis*. The *offspring* of female flies in which the mini-*white* gene had been de-repressed by a heat shock during early embryogenesis showed significantly increased eye coloration. This was seen even when crosses were constructed so that offspring did not have the GAL4 construct (i.e. it could not be explained by the continued, low-level production of GAL4). Thus, the induced epigenetic change could be retained through the female germ line and passed on to the next generation. Transmission through the male germ line was not observed.

The inheritance of epigenetic changes, in the guise of 'inheritance of acquired characteristics', has a long, controversial and largely undistinguished history. The results outlined above should not be taken as providing any sort of support for the possibility that phenotypic changes, such as shortened tails or strengthened muscles, can be passed on to the offspring. However, they do show that information other than that encoded by DNA sequences can be passed from parent to offspring and that this information is likely to be in the form of changes in chromatin structure or composition. Failure to find transmission through the male germ line is consistent with the extreme compaction of chromatin and loss of histones that occurs in sperm, something that would be expected to erase chromatin-based signals. The molecular basis of the chromatin signal remains to be determined, although some experiments suggest that histone H4 acetylation may be involved.

Imprinted genes

Perhaps the best examples of epigenetics in action have come from studies of genetic imprinting, the phenomenon whereby the activity of certain genes depends on whether they were inherited from the mother (via the ovum) or the father (via the sperm). The first imprinted gene (encoding the insulin-like growth factor *Igf2*) was identified and characterized only in 1991, but the existence of such genes was suspected well before then. There were several reasons for such suspicions, one of which was the finding that mouse embryos grown from zygotes prepared *in vitro* and containing either two copies of the maternal genome (gynogenotes) or two copies of the paternal genome (androgenotes) showed gross developmental abnormalities. Zygotes subjected to the same degree of *in vitro* manipulation, but containing one copy of each parental genome, developed perfectly well. It seems that the maternal and paternal genomes are not equivalent (despite having the same complement of autosomal genes) and that one copy of each is necessary for normal development.

Some imprinted genes are expressed only from the maternal allele and some from the paternal (Fig. 11.12). Interestingly, the latter often seem to be growth promoting genes (e.g. *Igf2*) whereas the former have the opposite effect (e.g. *Igf2r*, which encodes a receptor for the IGF2 protein that, unlike most receptors, simply sequesters the growth factor and prevents it exerting any effect on cell growth). This is consistent with the properties of androgenetic and gynogenetic embryos, which show increased and decreased growth, respectively. It has been

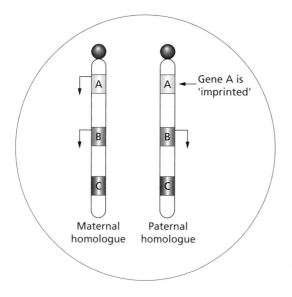

Fig. 11.12 An imprinted gene. In the hypothetical example shown, only the copy of gene *A* inherited from the mother (i.e. the copy on the maternal homologue of the chromosome carrying the gene) is expressed. The copy inherited from the father (on the paternal homologue) is silent. By convention, gene *A* is described as *paternally* imprinted. For some imprinted genes, the maternal allele is silent and the paternal allele is expressed and the gene is said to be *maternally* imprinted.

suggested that imprinted genes may have arisen, during mammalian evolution, as a result of the differing selection priorities of the maternal and paternal genomes. Survival of the paternal genome may benefit from a large embryo, based on the proposition that larger offspring are more likely to survive and reproduce. Conversely, although the maternal genome too will benefit from improvements in the reproductive fitness of the offspring, survival of the maternal genome (in the form of the mother herself) may be compromised if the size of the embryo is such as to make unacceptable nutritional or physical demands. Whatever their evolutionary origins, it seems that imprinted genes are largely a mammalian phenomenon. They are not seen in two of the geneticist's favourite organisms, the fruitfly *Drosophila* and the nematode worm *Caenorhabditis elegans*.

More than 20 imprinted genes are now known and of these, equivalent genes are present in both humans and mice and show the same pattern of imprinting. Examples are listed in Table 11.2. It can be estimated that there are probably about 500 imprinted genes in the human genome, less then 2% of the total number of genes, but still a significant minority. Their importance lies not just in their growth regulating properties, or in the fact that abnormal imprinting is found associated with many human diseases, including some cancers, but also in the fact that they pose a fundamentally new question concerning gene regulation: how does the cell impose a different level of transcriptional activity on two alleles that are, at the DNA level, identical?

The imprinting cycle

Imprints, once established, do not last forever. They are not passed on intact

Table 11.2 Examples of imprinted genes.

Gene*	Expressed allele	Chromosome location		Product and function
		mouse	human	
Igf2	p	7	11p15	Foetal growth factor
H19	m	7	11p15	Non-translated RNA
Ins2	p	7	11p15	Growth factor
Mash2	m	7	11p15	Placental transcription factor
p57^{kip2}	m	7	11p15	Cyclin kinase inhibitor
Snrpn	p	7	15q11-q13	RNA-processing protein
IPW	p	–	15q11-q13	Non-translated RNA
Znf127	p	7	15q11-q13	Zinc finger protein
Ube3a	m	7	15q11-q13	Ubiquitin ligase
Igf2r	m (bi-allelic)	17	6q	Igf2 binding protein and mannose-6-P receptor

*Genes are named using the mouse nomenclature, i.e. with lower case letters, with the exception of IPW, which has yet to be identified in the mouse.

from one generation to the next, but are instead erased during the early development of the male and female germ cells and then reset prior to germ cell maturation. We know this because some imprinted genes have been shown to be expressed from both maternal and paternal alleles (bi-allelically expressed) in primordial germ cells and also because high levels of DNA methylation associated with imprinting (see below) are also lost in these cells. This developmental cycle of imprint setting and erasure is shown in Fig. 11.13. The imprint is established during the later stages of germ cell maturation, with imprinted genes being marked 'maternal' or 'paternal'. In practice, only one mark is required, with its presence (indicated as a filled chromosome domain in the diagram) indicating that the gene has originated from one parent and its absence indicating the other. In the developing embryo, imprinted genes are regulated by promoters and enhancers, just as other genes are, and many show tissue-specific and stage-specific patterns of expression. But for all imprinted genes, in at least some tissues and at some developmental stages, one allele will be expressed and one will be silenced.

As increasing numbers of imprinted genes have been identified and characterized, it has become clear that they have a tendency to cluster in the genome. In humans, major clusters are found on the short arm of chromosome 11 (11p15) and the long arm of chromosome 15 (15q11-q13). Homologous imprinted genes are found in mice in two separate clusters on chromosome 7.

The Igf2/H19 cluster

The layout of the cluster of imprinted genes containing *Igf2* and *H19* and

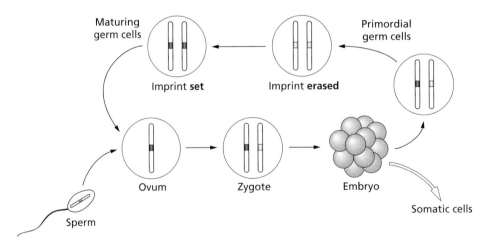

Fig. 11.13 The imprinting cycle. Existing imprints are erased early in germ cell maturation and new imprints appropriate to the type of germ cell (sperm or ovum) are put in place at a later stage. The diagram uses a female embryo, but the same cycle occurs in males. The molecular nature of the imprint remains obscure. CpG methylation is certainly a major component, but does not provide a complete explanation.

located at the distal region of mouse chromosome 7 is shown in Fig. 11.14. A similar cluster is located at 11p15.5 in humans. The cluster, like that at 15q11-q13 (see below) contains both maternally and paternally imprinted genes, intermingled with genes that are not imprinted. So the mechanisms that control imprinting must be able to operate locally, possibly on a gene-by-gene basis. It also shows properties that seem to be characteristic of imprinted genes (although certainly not exclusive to them), namely differentially methylated regions, repeat elements and noncoding (sometimes antisense) RNA transcripts.

The two genes that have been most closely studied within this cluster are *Igf2* and *H19*. *Igf2* is expressed only from the paternal homologue and encodes a foetal growth factor. Over-expression of *Igf2* during embryogenesis is associated with foetal overgrowth in both humans and mice. In contrast, *H19* is maternally expressed and produces a nontranslated RNA of unknown function. The *Igf2/H19* region contains two crucial control elements, an endoderm-specific enhancer (E) about 10 kb 3' of *H19* and an imprinting control region (ICR, so named for reasons that will soon become apparent) a few kb upstream of the *H19* promoter.

The function of these two regions has been assessed by construction of mouse strains in which they have been deleted ('knocked out') or translocated. Targeted deletion of the ICR results in bi-allelic expression of both *Igf2* and *H19*, i.e. loss of imprinting (Fig. 11.15b). Earlier experiments in which the whole of the *H19* locus was knocked out gave the same result, but only because the ICR had also been deleted. Animals that inherit the ICR deletion from their mothers are about 25% heavier than normal, consistent with increased (i.e. biallelic)

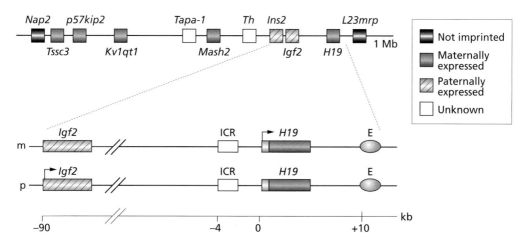

Fig. 11.14 The *Igf2/H19* gene pair is part of a cluster of imprinted genes on mouse chromosome 7. The upper part of the figure shows the genes in the cluster and their imprinting status. The lower part shows the *Igf2/H19* region in more detail. The imprinting control region (ICR) and enhancer (E) elements are shown.

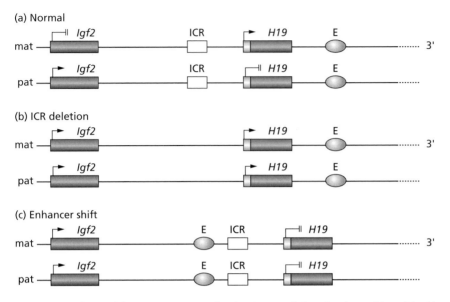

Fig. 11.15 Deletion of the imprinting control region (ICR) and changing the position of the 3′ enhancer (E) both cause loss of imprinting of the *Igf2* and *H19* genes.

expression of the *Igf2* growth factor. Paternal inheritance of the same deletion has no detectable phenotypic effect.

The chromatin structure of the ICR is different on the maternal and paternal chromosomes. The maternal allele has two prominent DNase hypersensitive sites (DHS) and is unmethylated. In contrast, the paternal allele is nuclease insensitive and is methylated at numerous CpGs (Fig. 11.16). Methylation spreads into the *H19* promoter and 5′ coding region and helps suppress *H19* transcription on the paternal chromosome.

A switchable boundary element

Chromosomes in which the enhancer element has been moved from its normal position and relocated just downstream of *Igf2* show expression of *Igf2* on the maternal chromosome (confirming the proximity effect), but not *H19* (Fig. 11.15c). In transgenic animals with the relocated enhancer, expression of *Igf2* showed the expected tissue distribution, i.e. it was strongly expressed in endodermally derived tissues such as liver, but not in nonendodermal tissues such as muscle or gut. The lack of *H19* transcription on the maternal chromosome when the enhancer is upstream is, at first sight, puzzling. We know that the promoter must be transcriptionally competent because it is normally active on the maternal chromosome. Why has moving the enhancer upstream caused it to remain inactive? The answer is that the ICR, in addition to its silencing properties, can also act as a boundary element, insulating one region of the chromosome from another.

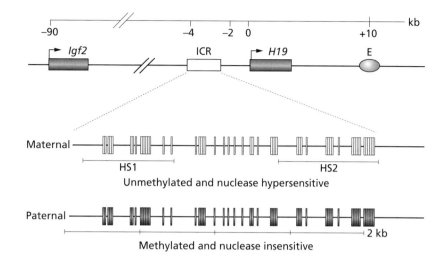

Fig. 11.16 The *Igf2/H19* imprinting control regions (ICR) on the maternal and paternal chromosomes differ in CpG methylation and nuclease sensitivity. The vertical lines across the maternal and paternal ICR in the lower part of the figure show CpG dimers that are either methylated (dark) or unmethylated (light). HS1 and HS2 are DNase hypersensitive sites. (From Bell & Felsenfeld, 2000.)

As discussed earlier, boundary elements such as Fab7 prevent DNA elements from exerting functional effects beyond their designated 'sphere of influence', possibly by limiting the spread of changes in chromatin structure from one DNA domain to the next. A crucial new property of the *H19* ICR boundary element is that it is *switchable*. On the *maternal* chromosome it exerts its boundary function and prevents the interaction between *Igf2* and the enhancer 3' of *H19*. But on the *paternal* chromosome it no longer acts as a boundary, so the *Igf2* promoter can interact with the 3' enhancer and the gene is expressed. The switch that determines whether or not the ICR can act as a boundary element operates through DNA methylation and a DNA-binding protein called CTCF.

On the maternal chromosome, the ICR is unmethylated and nuclease hypersensitive (i.e. its chromatin is accessible). GC-rich DNA elements within the ICR contain consensus sequences that can bind a protein called CTCF. This protein has been implicated in silencing of several mammalian genes. *In vitro* binding studies have shown that CTCF can bind efficiently to DNA fragments taken from the ICR and containing the appropriate consensus sequences, but the most exciting observation is that it binds only weakly to the same fragments when they are methylated at CpG elements. This observation provides a ready explanation (summarized in Fig. 11.17) for the switchable boundary function of the ICR and the expression patterns of the *H19* and *Igf2* genes. On the *paternal* chromosome, the ICR is methylated, CTCF cannot bind and the 3' enhancer can interact with the *Igf2* promoter, initiating transcription. On the same chromosome, methylation of the *H19* promoter suppresses its interaction with the en-

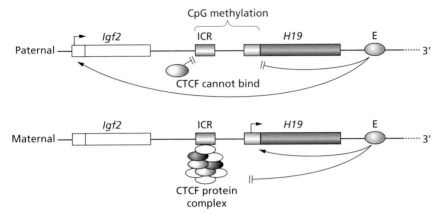

Fig. 11.17 The *Igf2/H19* imprinting control region (ICR) acts as a switchable boundary element that regulates interaction between the *Igf2* gene and the 3′ endodermal enhancer (E). (Redrawn from Bell & Felsenfeld, 2000.)

hancer. On the *maternal* chromosome, the ICR is unmethylated and accessible, CTCF can bind and initiate the formation of a boundary complex that blocks interaction between the enhancer and the *Igf2* promoter. The *H19* promoter is unmethylated and can interact, leading to *H19* transcription.

Deletion of a 1.2 kb region of the ICR leads to biallelic expression of *H19* in the mouse (although the differential DNA methylation of the maternal and paternal alleles remains unaltered). Remarkably, this same DNA element can bring about silencing in *Drosophila* when assembled into a test construct of the kind described earlier for studying Fab-7 silencing (Fig. 11.10). When it was inserted between the UAS and the *LacZ* and *white* promoters, it prevented their activation by GAL4. The mechanism by which this small piece of mouse DNA can induce genetic silencing in such an evolutionarily distant organism (in which DNA methylation is minimal) remains a mystery.

The situation in the *Igf2/H19* region, in which a single enhancer element 'chooses' between genes, is reminiscent of the regulation of expression of the embryonic, foetal and adult β-globin genes by the 5′ LCR (Chapter 6). The physical problems are much the same as well. The *Igf2* promoter is about 100 kb (or 5000 nucleosomes) away from the enhancer. Models in which a chromatin conformation 'spreads' from the enhancer to the promoter are unsatisfactory because of the lack of expression (on the paternal chromosome) of the *H19* gene. Looping models are often used for convenience, but we have little or no evidence that anything of the sort actually occurs.

The 15q11–q13 cluster

It has been known for some time that chromosome deletions that remove all or part of this region can cause two clinically distinct birth defects, depending on whether they are inherited from the mother or the father. Large deletions (at least 5 Mb) within the q11–q13 region of the *paternal* chromosome 15 lead to Prader–Willi Syndrome (PWS). The disease is rare (1:25 000 births) and is characterized by obesity, failure to thrive in infancy and a variable degree of mental retardation. Conversely, large deletions of the *maternal* copy of 15q11–q13 lead to Angelman Syndrome (AS). This occurs with a similar frequency to PWS but the symptoms are quite different. AS is characterized by severe mental retardation, seizures, hyperactivity and a characteristic 'puppet-like' walk and jerky arm movements. In most cases of PWS the deletions in the paternal 15 arise *de novo* (i.e. they are not present in the father himself). A clue that imprinting was involved came from the finding that PWS patients, with no detectable deletions of 15q11–q13, often had two copies of the *maternal* 15 (i.e. the paternal 15 was missing altogether). Also, a small number of AS patients without detectable deletions have been found to have two *paternal* copies of chromosome 15. How this situation, known as *uniparental disomy*, might arise is shown in Fig. 11.18.

It is difficult to develop imprinting theories based on the effects of relatively large chromosome deletions, or of uniparental disomies, because large numbers of genes, sometimes imprinted in opposite directions, are inevitably involved. Fortunately, by more detailed analysis of small deletions in some PWS patients it was possible to define a region of about 1000 kb that was of particular

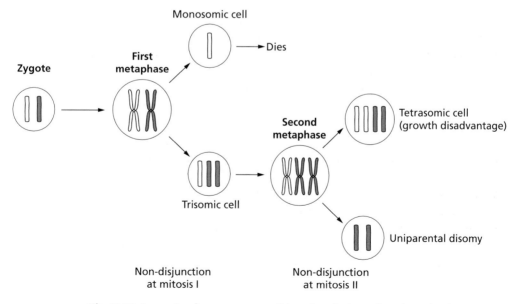

Fig. 11.18 Successive chromosome non-disjunctions during early embryonic mitoses can give rise to uniparental disomy for specific chromosomes. Maternal and paternal homologues are shown as light and dark, respectively. When this type of event involves chromosomes carrying imprinted genes, the resulting mouse strains can provide valuable information on how imprinting control mechanisms operate.

importance. This region has been cloned as a yeast artificial chromosome (YAC, a system that allows cloning of much larger stretches of DNA than bacterial and viral vectors). The region contains several genes, some of which are expressed only from the paternal chromosome, and just one (*UBE3A*) that is expressed maternally (Fig. 11.19). On the maternal chromosome, this region is more highly methylated at CpG-rich sites (see below) and replicates later in S-phase. In other words, it has the properties of genetically silent heterochromatin on the maternal chromosome and of euchromatin on the paternal one. Perhaps PWS results from loss of expression of some or all of the paternally expressed genes in the cluster, while AS results from loss of *UBE3A* expression (Fig. 11.19). Support for this possibility comes from studies on two AS patients who have neither detectable chromosome deletions nor uniparental disomy, but who do have mutations in the *UBE3A* gene. *UBE3A* encodes a ubiquitin ligase, an enzyme used to label proteins destined for intracellular degradation through the proteosome pathway.

The PWS/AS region provides a fascinating illustration of how gene silencing may operate over large distances in a 'real' cellular environment (i.e. not in transgenic animals or artificial DNA constructs). Several PWS and AS patients have been identified who have small deletions (microdeletions) within two very tightly defined regions within this larger domain. Because these regions have such far-reaching effects on imprinting, they can be designated imprinting control regions (ICRs). Their locations are shown in Fig. 11.19. They all occur within the coding or upstream control regions of a paternally expressed gene *SNRPN*, encoding a small nuclear ribonuclear protein. The gene has a complex structure, with large numbers of exons and several promoters. Intriguingly, the microdeletions that cause AS lie within a quite different region of the gene from

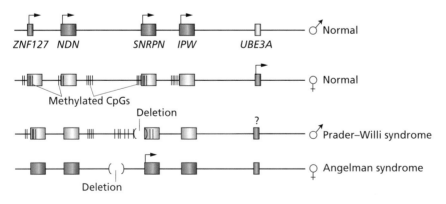

Fig. 11.19 The Angelman/Prader–Willi imprinted region on human chromosome 15. Normally, the genes shown are expressed either paternally (dark boxes) or maternally (light boxes). Small deletions within the imprinting control region upstream of the *SNRPN* gene can alter CpG methylation and lead to inappropriate patterns of gene expression. (Redrawn from Tilghman, 1999.)

those that cause PWS. On *paternal* chromosomes carrying the *PWS microdeletions*, there is increased CpG methylation and lack of gene expression across the whole 1–2 Mb chromosome domain, i.e. the domain has the properties that it should have if it were on the maternal chromosome. Conversely, on *maternal* chromosomes carrying the *AS deletions* there is lack of CpG methylation and expression of genes that would normally be silent (Fig. 11.19). Note that if a PWS deletion is inherited from the mother (i.e. is on the maternal homologue), or if an AS deletion is inherited from the father (i.e. is on the paternal homologue), they do not affect the normal inactive or active state of the domain.

The mutations identified in PWS and AS patients have provided results consistent with much work done with transgenic animals and artificial DNA constructs. They show that small DNA elements, sometimes designated imprinting centres, are able to influence chromatin structure, and thereby gene expression, over very large chromosome domains. They also confirm that genes within the same imprinted domain can be imprinted in different directions, suggesting that some elements of the regulatory processes involved must operate on a gene-by-gene basis. It is interesting to remember that different genes can respond to the same chromatin environment in very different ways. As discussed in Chapter 9, some genes are transcriptionally active only in a heterochromatin environment and become inactive when translocated into euchromatin. So, an imprinting process that leads to a consistent change in chromatin structure across the whole gene cluster could still result in genes within that cluster being imprinted in opposite directions.

DNA methylation as an epigenetic marker

The methylation of selected cytosine residues in DNA would seem, at first sight, to be the ideal epigenetic modification. The altered base, 5-methylcytosine, is a stable component of the DNA helix and is about as close as one can get to altering the DNA sequence without actually doing so. Cytosine methylation does not alter the coding properties of DNA, but it does influence the structure of chromatin, usually in such a way that transcription is inhibited. The structural effects of methylation are probably mediated by proteins that specifically recognize methyl-cytosine residues, either singly or in clusters. These mechanisms have been discussed previously (Chapter 10).

The circumstantial evidence for the involvement of cytosine methylation in gene silencing, in general, and imprinting, in particular, is very strong. It is further supported by the finding that many parental imprints are lost in embryonic stem cells from mice lacking DNA methyltransferase activity (dnmt-/-). Despite this, methylation alone is unlikely to provide a complete explanation for parental imprinting. One reason is that the genome of the developing embryo undergoes a massive, generalized demethylation between fertilization and the preimplantation blastocyst stage. Most of the differentially methylated regions

(DMRs) that distinguish maternal and paternal alleles are caught up in this wave of demethylation, with the result that the great majority of parent-specific methylation patterns are erased. They are re-established postimplantation. However, some isolated regions survive these changes. Importantly, the ICR upstream of *H19* is one such region. This region is methylated in sperm and undermethylated in the egg. It is demethylated postfertilization, with the exception of a 2 kb core element that retains its sperm-specific (paternal) methylation. Later in development, the whole region is remethylated on the paternal allele. This 2 kb element is just upstream of the cluster of direct repeats and includes the silencer element that was shown to operate in *Drosophila*. These findings suggest that, for some genes, the imprint may be carried by small, differentially methylated regions that are protected in some way from generalized demethylation early in development.

Imprinting may require various mechanisms of silencing and activation

Imprinting involves a hierarchy of different mechanisms that lead from the imprint that distinguishes the maternal and paternal homologues, to the chromatin structures that determine transcriptionally silent or active states. For the 'primary' imprint (i.e. the epigenetic mark that passes through the germ line) the best candidate so far is DNA methylation. But, as noted above, it is not a complete explanation. Histone acetylation, and even specific chromatin structures, are possible mechanisms for transmission through the female germ line, but the extreme compaction and histone loss that occurs in sperm chromatin makes them unlikely candidates for the male germ line. The mechanisms that link the primary imprint with the final chromatin state will differ in detail from one gene to another, but are likely to include many of the processes for initiation and maintenance of active and silent states that have been dealt with in this and previous chapters.

One characteristic of imprinted gene clusters that has not been touched upon so far is their tendency to include noncoding RNA transcripts. Sometimes noncoding RNAs have a defined function in determining chromatin structure and transcription. In the next chapter, we will discuss the *Xist* and *roX* RNAs that play central roles in determining the transcriptional activity of X chromosomes in mammals and *Drosophila*, respectively. The function of the *H19* RNA is not known, nor are the functions, if any, of the antisense transcripts that crop up in many imprinted regions. One interesting possibility is that, for these regions, it is the act of transcription itself that is important, rather than the transcript. Antisense transcripts on the maternal *Igf2* allele may help suppress its transcription in the sense direction. Conversely, the presence of functional transcription machinery in a chromosome domain, even when operating at low levels, may be enough to maintain an open chromatin conformation that is permissive for the up-regulation of other genes within the domain when conditions are right. Developmentally regulated transcription has been detected in intergenic, noncoding DNA along the β-globin domain, and may fill a similar role. Perhaps, in this case, transcription prevents chromatin condensation in intergenic regions, something that might bring about the inappropriate silencing of adjacent, active genes.

Further reading

Epigenetics

Jablonka, E. & Lamb, M. J. (1995) *Epigenetic Inheritance and Evolution*. Oxford University Press, Oxford.
Jenuwein, T. & Allis, C. D. (2001) Translating the histone code. *Science*, **293:** 1074–1079.
Rideout, W. M., Eggan, K. & Jaenisch, R. (2001) Nuclear cloning and epigenetic reprogramming of the genome. *Science*, **293:** 1093–1098 (*see also related review articles in this issue*).
Turner, B. M. (2000) Histone acetylation and an epigenetic code. *BioEssays*, **22:** 836–845.

Drosophila development and homeotic genes

Lawrence, P. A. (1992) *The Making of a Fly*. Blackwell Scientific Publications, Oxford.
McGinnis, W. & Krumlauf, R. (1992) Homeobox genes and axial patterning. *Cell*, **68:** 283–302.

Polycomb-induced silencing

Breiling, A., Turner, B. M., Bianchi, M. & Orlando, V. (2001) General transcription factors bind promoters repressed by Polycomb group proteins. *Nature*, **412:** 651–655.
Birchler, J. A., Bhadra, M. P. & Bhadra, U. (2000) Making noise about silence: repression of repeated genes in animals. *Curr. Opin. Genet. Devel.*, **10:** 211–216.
Cavalli, G. & Paro, R. (1998) Chromo-domain proteins: linking chromatin structure to epigenetic regulation. *Curr. Opin. Cell Biol.*, **10:** 354–360.
Cavalli, G. & Paro, R. (1998) The *Drosophila* Fab-7 chromosomal element conveys epigenetic inheritance during mitosis and meiosis. *Cell*, **93:** 505–518.
Gerasimova, T. I. & Corces, V. G. (1998) Polycomb and trithorax group proteins mediate the function of a chromatin insulator. *Cell*, **92:** 511–521.
Pirrotta, V. (1997) PcG complexes and chromatin silencing. *Curr. Opin. Genet. Dev.*, **7:** 249–258.

Imprinting

Allshire, R. & Bickmore, W. (2000) Pausing for thought on the boundaries of imprinting. *Cell*, **102:** 705–708.
Gregory, R. I., Randall, T. E., Johnson, C. A. *et al.* (2001) DNA methylation is linked to deacetylation of histone H3, but not H4, on the imprinted genes *Snrpn* and *U2af1-rs1*. *Mol. Cell Biol.*, **21:** 5426–5436.
John, R. M. & Surani, M. A. (1996) Imprinted genes and regulation of gene expression by epigenetic inheritance. *Curr. Opin. Cell Biol.*, **8:** 348–353.
Kelsey, G. & Reik, W. (1997) Imprint switch mechanism indicated by mutations in Prader–Willi and Angelman syndromes. *BioEssays*, **19:** 361–365.
Killian, J. K., Byrd, J. C., Jirtle, J. V. *et al.* (2000) *M6P/IGF2R* imprinting evolution in mammals. *Mol. Cell*, **5:** 707–716.
Reik, W. & Walter, J. (1998) Imprinting mechanisms in mammals. *Curr. Opin. Genet. Dev.*, **8:** 154–164.

Tilghman, S. M. (1999) The sins of the fathers and mothers: genomic imprinting in mammalian development. *Cell*, **96:** 185–193.

Igf2/H19 and imprinting mechanisms

Bell, A. C. & Felsenfeld, G. (2000) Methylation of a CTCF-dependent boundary controls imprinted expression of the *Igf2* gene. *Nature*, **405:** 482–485.

Reik, W. & Murrell, A. (2000) Silence across the border. *Nature*, **405:** 408–409.

Srivastava, M., Hsieh, S., Grinberg, A. *et al.* (2000) *H19* and *Igf2* monoallelic expression is regulated in two distinct ways by a shared *cis* acting regulatory region upstream of *H19*. *Genes Dev.*, **14:** 1186–1195.

Noncoding RNA and antisense transcripts

Barlow, D. P. (1997) Competition—a common motif for the imprinting mechanism? *EMBO J.*, **16:** 6899–6905.

Matzke, M., Matzke, A. J. M. & Kooter, J. M. (2001) RNA: guiding gene silencing. *Science*, **293:** 1080–1083.

Moore, T., Constancia, M., Zubair, M. *et al.* (1997) Multiple imprinted sense and antisense transcripts, differential methylation and tandem repeats in a putative imprinting control region upstream of mouse *Igf2*. *Proc. Natl. Acad. Sci., USA*, **94:** 12509–12514.

Chapter 12: Mechanisms of Dosage Compensation

Introduction

Sexual reproduction is extremely common among eukaryotes. Even single-celled eukaryotes such as yeasts, which normally multiply by simple cell division, have a mechanism whereby cells can adopt one of two mating types and then conjugate to form diploids (see Chapter 10). Likewise plants, which can usually spread perfectly well by sending out shoots or runners, often have an alternative, sexual mode of reproduction. But sex is complicated, requiring, in higher eukaryotes at least, developmental pathways that lead to male and female sexual organs and the physiological and biochemical apparatus required for meiosis, germ cell maturation, the attraction of partners and mating. The high cost in terms of complexity must be offset, in evolutionary terms, by the advantages conferred by the enormous increase in genetic variability that sexual reproduction brings. The reshuffling of alleles that occurs with every sexual generation will produce a population able to cope with environmental shifts more effectively than a relatively homogeneous population derived from asexual methods of reproduction. The crucial point, and the definitive measure of evolutionary success, is that variable populations are better able to avoid the ultimate catastrophe of extinction.

Methods of sex determination

Genetic mechanisms for defining different sexes vary widely from one organism to another. The simplest systems involve a single locus that is homozygous in one sex, the *homogametic* sex, and heterozygous in the other, the *heterogametic* sex (Fig. 12.1). This simple system has evolved in different ways to reach varying levels of complexity in different organisms. Mechanisms have been put in place to suppress meiotic recombination (crossing over) of the sex-determining alleles in the heterogametic sex in order to prevent the generation of mixtures of alleles leading to intersex states. The inability to recombine can spread to include part or all of one chromosome, with an accompanying loss of genetic information (Fig. 12.1). The evolutionary pressures that have driven this chromosome degeneration are still not understood, but the end result in many species is that the two sexes show differences not just in alleles at one or a few loci, but in complete chromosomes. In most species, including our own, it is the males who carry the degenerate chromosome, although there are many exceptions (birds and lizards, for example).

Before moving on to the main topic of this chapter, it is interesting to ask

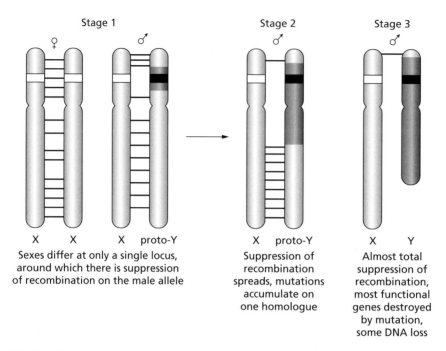

Fig. 12.1 How sex chromosomes might have evolved.

why it is that chromosomal differences between the two sexes have evolved so frequently. Sexual differentiation, as we shall see, is usually triggered by the switching on or off during development of one or a small number of crucial switch genes. The products of these genes initiate a cascade of gene regulatory events that mediate progression down one or the other pathway of sex determination. This is exactly what happens in mammals. In humans, for example, it is the protein product of the *SRY* gene on the Y chromosome that sends the early embryo down the male pathway of sexual development. One does not need major chromosome differences to operate a mechanism of this sort. So why have they arisen so often? It may be that, for reasons we don't yet understand, chromosome differences between the sexes are crucially important and necessary for the development of complex, multicellular organisms. However, it is worth bearing in mind that they may have occurred as a by-product of the suppression of crossing-over. Mathematical analysis of the factors that influence the spread of alleles through populations shows that suppression of crossing-over will lead *inevitably* to the gradual accumulation of deleterious mutations (perhaps even of mutations that cause further suppression of crossing-over). This will lead to the progressive degeneration of one of the two originally homologous chromosomes (Fig. 12.1). There is no *selection* for this eventuality; it just happens as a consequence of the initial step of adopting a two-sex strategy for reproduction. But, whatever the evolutionary drive behind chromosome degeneration, the fact that it has occurred (and is presumably continuing) has required the co-evolution of mechanisms to cope with major chromosomal differences between members of the same species.

Chromosomal methods of sex determination create a need for dosage compensation

In both mammals and *Drosophila*, males have one copy of each of two different sex chromosomes, X and Y, while females have two copies of the X. In both groups of organisms, the Y is gene-poor and largely heterochromatic. It contains just a few genes needed for male development or fertility. In contrast, the X is a large, gene-rich chromosome (Fig. 12.2). A twofold difference in copy number, if left uncorrected, would result in a twofold difference between the sexes in the intracellular concentrations of several thousand gene products. It is not surprising that evolution has been unable to accommodate such a massive difference between members of the same species. It has instead developed ways of eliminating it through mechanisms of dosage compensation.

There are three conceptually straightforward ways in which levels of X-linked gene transcripts could be equalized between the heterogametic and homogametic sexes. These are: (i) switching off genes on one of the two female Xs; (ii) doubling the rate of transcription of genes on the single male X; and (iii) halving the rate of transcription on each of the two female Xs. The end result in both (ii) and (iii) is the same, in that genes on the single male X are transcribed twice as fast as the equivalent genes on the two female Xs, but the routes used to reach this state are fundamentally different. Examples of all three mechanisms have been observed. Mammals use the first, *Drosophila* the second and the nematode worm *Caenorhabditis elegans* the third. The fact that three very different mechanisms of dosage compensation have evolved, apparently independently, confirms the importance of the final objective, namely equalizing the levels of X-linked gene products between the sexes. In all three groups of organisms, mutations that disrupt dosage compensation are lethal in the affected sex.

Dosage compensation presents the cell with gene regulation problems that do not arise in quite the same way in other circumstances and that we have not encountered before. First, it requires that genes along a whole chromosome should be regulated in the same way. In other words, a chromosome-wide control must be exerted that overrides local control mechanisms operating on individual genes. Second, in the case of the mammalian mechanism, the cell must 'choose' between genetically identical chromosomes in the same nucleus, inactivating one but not the other. Third, in the *Drosophila* and *C. elegans* mechanisms, no choice is necessary, but the regulation must be *quantitative*, producing a twofold up- or down-regulation, respectively. In what follows we will deal in turn with each of the three different types of dosage compensation mechanisms. There are many unanswered questions, but it is now clear that changes in chromatin structure are fundamentally important in each mechanism.

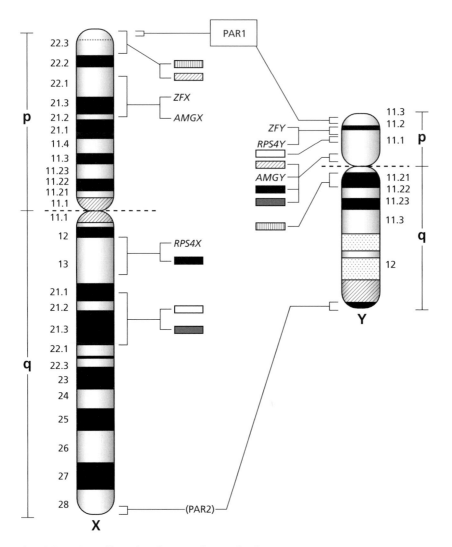

Fig. 12.2 Regions of homology between the X and Y chromosomes suggest a common evolutionary origin. 2.6 Mb of DNA at the extreme distal ends of the short arms is common to both the X and Y chromosomes. Known as the pseudoautosomal region (PAR1), it is the site at which the X and Y chromosomes associate during male meiosis, is essential for completion of meiosis in males and has an exceptionally high frequency of recombination. There is a minor, nonessential PAR at the tip of the long arms. Three examples of genes that have expressed homologues on both the X and Y are shown (*ZFX/Y*, *AMGX/Y* and *RPS4X/Y*). Also shown (as boxes with different types of shading), are blocks of non-coding DNA that occur on both chromosomes. Note that the *distribution* of these homologous sequences is very different between the two chromosomes, indicative of frequent chromosome rearrangements during evolution. G-bands are shown in black and regions of heterochromatin are dotted or cross-hatched. (Redrawn from Strachan & Read 1996.)

Dosage compensation in mammals

The heterochromatic structure of the inactive X

In mammals, most of the genes on one of the two X chromosomes in female cells are inactivated early in development, at around the time of implantation. Once inactivation of one X has occurred, then that X (designated Xi) remains inactive through all subsequent cell generations. X inactivation is a particularly stable epigenetic modification and experiments attempting to reverse the silent state once it is in place have met with only limited success.

Xi shares several chemical and structural properties with constitutive heterochromatin, and is said to consist of *facultative* heterochromatin. It replicates in the latter part of S-phase in many cells, has relatively high methylation of selected CpG residues, and shows distinctive histochemical staining in both interphase and metaphase cells. Like some types of constitutive heterochromatin, Xi often remains visible through interphase, forming a distinctive structure, the Barr Body, usually located adjacent to the nuclear envelope (Fig. 12.3). The chromatin within this structure is often described as 'condensed', but careful microscopical analysis and three-dimensional reconstruction of Xa and Xi chromosomes labelled with X-specific DNA probes suggest that the difference between them is more a matter of shape than of the amount of chromatin per unit volume. The properties of the chromatin of Xi are summarized in Table 12.1.

A further parallel between Xi and constitutive heterochromatin has come from the use of indirect immunofluorescence microscopy to study the level of histone acetylation along metaphase chromosomes. In human and murine cells, both constitutive, pericentric heterochromatin and the inactive X label only weakly with antibodies to the acetylated isoforms of H2A, H2B, H3 and H4 (an example is shown in Plate 3.2). As discussed in Chapter 8, histone under-

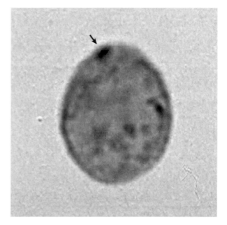

Fig. 12.3 The Barr Body in a human peripheral lymphocyte (arrow). The facultative heterochromatin of the inactive X chromosome in female cells can be seen under the light microscope after fixation and staining with DNA binding dyes such as Giemsa. The structure is known as the Barr Body, after its discoverer. (Photograph from Tessa Webb.)

Table 12.1 Properties of Xi in eutherian and marsupial mammals.

Property	Eutherian	Marsupial
Selective CpG methylation	Yes	No
Presence of XIST RNA	Yes	No
Homologue inactivated	Randomly chosen*	Paternal
Stability of inactive state	Hyperstable	Labile
Tissue dependence	Tissue-independent	Tissue-dependent
Sex chromatin body	Yes†	Occasional
Replication timing of Xi	Late	Late
H4 acetylation	Underacetylated	Underacetylated

*In cells of the inner cell mass that will form the embryo proper. In cells of the trophectoderm, forming the extra-embryonic tissue, Xi is always the paternal homologue.
†An example is shown in Fig. 12.3.

acetylation is a general property of heterochromatin. Xi is also enriched in a variant form of histone H2A, macro-H2A (Chapter 4), but this variant is not detected at higher than normal levels in centric heterochromatin.

Some genes escape inactivation

Not all genes on Xi are transcriptionally inactive. A cluster of genes in a region of the X known as the pseudo-autosomal region (often abbreviated to PAR) escape inactivation. Many of these genes are known to have homologues on the Y chromosome and the region pairs with the Y during male meiosis (Fig. 12.2). Such pairing is necessary for the successful completion of meiosis. Other genes scattered along the X also escape inactivation (Fig. 12.2). These sometimes have homologues on either the Y or the autosomes, reducing or eliminating the imbalance between males and females. Thus, although X inactivation is a chromosome-wide process, specific regions are protected from its effects. A similar situation has been observed in some cases of position effect variegation (PEV), where the spread of the inactive state can skip intervening regions leaving them transcriptionally active (see Chapter 10) or in some imprinted chromosome regions where adjacent genes can be oppositely imprinted (Chapter 11).

The X inactivation centre (*XIC*)

Studies of abnormal X chromosomes in humans and mice have shown that an X chromosome can be inactivated only if it contains a specific, defined region. Chromosomes that lack this region can never form an inactive X, whereas those that retain it can, even if other parts of the chromosome are missing. The region has been designated the X inactivation centre (abbreviated to *XIC* in humans and *Xic* in mice). In humans, analysis of the properties of large numbers of naturally occurring X–autosome translocations has located the *XIC* to a region about 400 kb in size that maps to Xq13 (Fig. 12.2). The equivalent locus in mice maps to band XD (Fig. 12.4).

Several genes have been located within the *XIC* (Fig. 12.4). Most of these are inactivated normally, but one is unique in that, in adult somatic cells, it is expressed *only* from Xi. The gene has been identified in both mice and humans and designated X inactive specific transcript (*XIST* in humans, *Xist* in mice). There is only 70% homology overall between the mouse and human genes, indicating that they are evolving rapidly. The *XIST/Xist* genes do not code for a protein product, have no conserved open reading frames and their transcripts are restricted to the cell nucleus. *In situ* hybridization has revealed that *Xist* RNA co-localizes with the Xi (the Barr Body) in interphase cells, and is said to 'coat' it.

It has recently been shown that an 'antisense' transcript is also produced by the *Xist* region, the product of a gene designated *Tsix*. (The use of 'sense' and 'antisense' is historical as neither transcript encodes a protein product.) The *Tsix* transcript begins in a region designated DXPas34 a few kb downstream of the last *Xist* exon and terminates about 10 kb upstream of the *Xist* promoters (Fig. 12.4). *Tsix* is transcribed only in the early stages of X inactivation and may have a role in determining which of the two Xs remains active. The role of the *Xist* and *Tsix* RNA transcripts is discussed further below.

The developmental biology of X inactivation

X inactivation occurs in the blastocyst at around the time of implantation and is

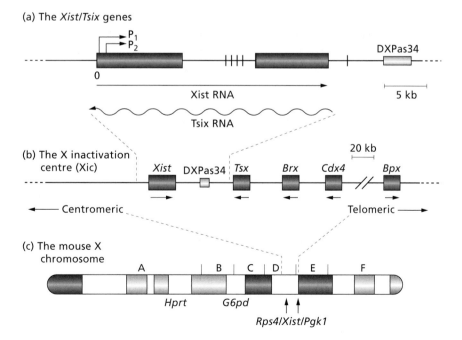

Fig. 12.4 The *Xist* and *Tsix* genes and the X inactivation centre (*Xic*) in the mouse.

probably complete by the onset of gastrulation, at 6.5 days *post coitum* (dpc) in mice. This is the developmental stage at which patterns of DNA methylation associated with tissue-specific gene expression begin to be put in place. In the epiblast cells of the inner cell mass (ICM), which will go on to form the embryo itself, the choice of which X to inactivate is essentially random (Fig. 12.5). Thus, on average, the maternal and paternal homologues will be inactivated in about 50% of cells. In any given animal, this distribution will vary, by chance, from one cell lineage to another. In mice or humans with more than two X chromosomes, then all but one are inactivated at this stage. (This is often referred to as the $n-1$ rule and the process by which it occurs as 'counting'.) Humans with three or even four X chromosomes are essentially normal, fertile females. The contrast between this very mild phenotype and the disastrous effects of autosomal trisomies demonstrates the effectiveness of the X inactivation process.

There must be a mechanism to ensure that one X chromosome, but no more than one, remains active in every cell. The way in which this is achieved remains unclear, but one attractive suggestion is that it involves a blocking factor that binds specifically to one X chromosome and prevents its inactivation. If this blocking factor is produced in such limiting amounts that only one X chromosome per cell can be protected (i.e. remains active), then this will account for the $n-1$ rule and the apparent ability of cells to 'count' their X chromosomes.

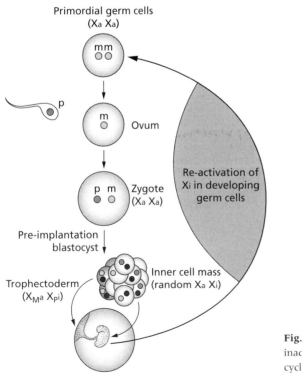

Fig. 12.5 The X-inactivation/reactivation cycle in female mammals.

The problem with this useful model is that the hypothetical blocking factor has not yet been identified.

A subgroup of cells within the ICM constitutes the trophectoderm and the primitive endoderm and will go on to form extraembryonic tissues such as the placenta and the amniotic membrane. In contrast to the random inactivation in the embryo proper, in these cells it is always the *paternal* X that is inactivated (Fig. 12.5). This has been shown by microscopical analysis of embryos in which the maternal and paternal X chromosomes are cytologically distinguishable. Thus, there must be an *imprint* that is read in a tissue-specific way on either the paternal X, to *facilitate* its inactivation in the trophectoderm, or the maternal X, to *prevent* its inactivation in the trophectoderm. (We return to this later.)

The stability of Xi and its resistance to reactivation have been emphasized. However, inactivation is reversed in the germ cell lineage in female embryos, so that in the premeiotic oocyte both X chromosomes are active. It has been suggested that reactivation is necessary in order to allow efficient pairing of the two X homologues during meiosis. The chromatin structure of Xi may not be appropriate either for pairing or for the breakage and rejoining of DNA required for meiotic crossing over. This possibility is consistent with the observation that the single X chromosome in premeiotic *male* germ cells shows, transiently, some of the properties of Xi, in the form of increased *Xist* transcript levels and gene silencing, though not reduced H4 acetylation. These changes could help prevent the single male X from engaging in inappropriate, and potentially lethal, crossing-over.

Xist, *Tsix* and the initiation of X inactivation

There is strong evidence that the *Xist* gene and its RNA product provide the switch that initiates X inactivation in *cis*. (i) *Xist* is unique in being expressed only from Xi; (ii) *Xist* RNA levels increase dramatically in preimplantation embryos at the time of X inactivation; (iii) *Xist* up-regulation precedes X inactivation and appears to be an absolute requirement for it to occur (X chromosomes carrying small deletions of the *Xist* promoter cannot be inactivated); (iv) *Xist* RNA co-localizes with Xi in interphase nuclei and, in mice, is distributed along one of the two metaphase X chromosomes; and (v) *Xist*-containing transgenes can induce at least some of the properties of inactive chromatin when inserted, in multiple copies, into autosomes.

Attempts to unravel the factors that throw the *Xist* switch during development have revealed an unexpected complication. The increase in *Xist* RNA in the late blastocyst is not the result simply of up-regulation of transcription: it transpires that *Xist* is transcribed from all X chromosomes in both male and female preimplantation embryos. However, the transcript is unstable, so levels of *Xist* RNA remain low. In the late blastocyst, the transcript is stabilized (by an unknown mechanism) resulting in an increase in the steady-state level of *Xist* RNA and coating one of the X chromosomes in *cis*.

258   CHAPTER 12

Although the evidence outlined above establishes the functional importance of *Xist* RNA, it begs the important question as to what causes the stabilization of *Xist* RNA and its spreading on one chromosome (in normal
circumstances) but not the other. In other words, how does the cell 'choose'
which X to inactivate and which to keep active? It seems that the choice stage
involves the *Tsix* 'antisense' transcript. In preimplantation female embryos,
both X chromosomes produce both *Xist* and *Tsix* transcripts. *In situ* hybridization
shows that both transcripts stay adjacent to their site of synthesis. Quantifica-
tion is difficult, but *Tsix* may be a more common transcript than *Xist* at this stage.
Later in development, *Tsix* transcription on *one* X is down-regulated. This is
followed by stabilization of the *Xist* transcript, increased levels of *Xist* RNA
and coating (and inactivation) of the chromosome in *cis* (Fig. 12.6). Thus, the
'choice' mechanism that determines which X to inactivate involves the con-
tinuation or down-regulation of *Tsix* transcription. Once the decision has been
made, *Tsix* transcription is lost from both Xa and Xi. Consistent with this model,
if one of the two X chromosomes carries a deletion of the *Tsix* promoter region
(i.e. does not express *Tsix*), then that chromosome is *always* chosen to be
inactivated.

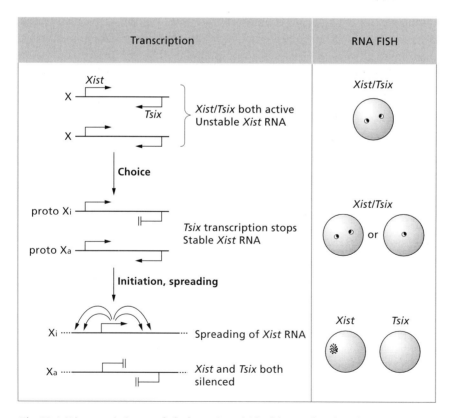

Fig. 12.6 *Tsix* transcription may help determine which of the two female X chromosomes
remains active during random X inactivation in the epiblast.

Tsix transcription plays a crucial role in imprinted X inactivation. Experiments with interspecies mouse crosses (in which the maternal and paternal *Xist* and *Tsix* transcripts could be distinguished) have shown that, in the early, preimplantation embryo, *Tsix* is expressed exclusively from the maternal allele and *Xist* from the paternal. This pattern of expression continues in the trophectoderm and primitive endoderm, with the result that the paternal X is always inactivated in the extraembryonic tissues. In contrast, in the epiblast cells of the ICM, differences in *Tsix* and *Xist* expression between the maternal and paternal chromosomes are erased, with the result that either is equally likely to be inactivated. The model is summarized in Fig. 12.7. It has been suggested that the differential expression of *Tsix* from the maternal and paternal chromosomes results from a germ-line-specific imprint. The imprint may lie in the CpG-rich region in which *Tsix* transcription initiates and that is known to be differentially methylated. Whatever its nature, the *Tsix* imprint is the primary determinant of X inactivation in the extraembryonic tissues. Extra copies of the *maternal* X chromosome are *not* inactivated in the extraembryonic tissues, while extra copies of the *paternal* X are always inactivated. Only in the epiblast, following erasure of the imprint, can the counting mechanism operate successfully.

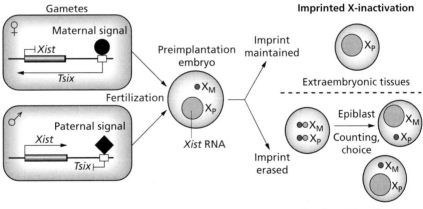

Fig. 12.7 A model for regulation of imprinted X inactivation. The model proposes that the CpG region around the transcription start site of the *Tsix* gene is the target of signals (imprints) put in place in the maternal and/or paternal germ cells. These signals lead to expression of *Tsix* and silencing of *Xist* in the *maternal* gametes and silencing of *Tsix* and expression of *Xist* in the *paternal* gametes. This carries through into the early embryo, with the result that the paternal X is inactivated. This situation is retained in the cells that make up the extra-embryonic tissues (placenta, etc.), but in the epiblast cells that go on to form the embryo proper, the imprints are erased and random X inactivation ensues through a counting/choice mechanism. Levels of *Xist* RNA and its spreading on Xi are indicated by large or small shaded discs. *Tsix* expression is indicated by small black discs. (Redrawn from Lee 2000.)

Xist RNA, gene silencing and heterochromatin assembly

The nature of heterochromatin has been discussed already in terms of its DNA and protein components. RNA has not featured. This is, at least in part, for technical reasons. The chemistry of RNA is such that the identification of specific RNA molecules and their structural and functional characterization is technically demanding. As RNA is a ubiquitous, common and extraordinarily heterogeneous component of the cell nucleus, it will be found to be associated with virtually every nuclear component, including heterochromatin. The great majority of these associations will be fortuitous and nonsignificant. However, in the case of the *Xist* transcript, we have a family of RNA molecules that associate, in *cis*, with a specific, genetically silent and largely heterochromatic chromosome. So, is *Xist* RNA a major determinant of the structure and/ or function of facultative heterochromatin? The balance of evidence suggests that it is.

First, the association of *Xist* RNA with Xi is selective. It is found neither along the pseudoautosomal region, which remains active and euchromatic, nor in association with constitutive (centric) heterochromatin. These observations show that *Xist* RNA does not simply coat nonspecifically any chromatin that happens to be in its vicinity. Second, when *Xist* transgenes are inserted in multiple copies into some mouse autosomes, their (over)expression results in coating of the autosome in *cis* and the adoption of a local chromatin structure that is late replicating, underacetylated and whose genes appear not to be transcribed. These findings suggest that *Xist* expression may be both necessary and sufficient to trigger the adoption of a heterochromatin-like structure. Perhaps surprisingly, continuing *Xist* expression is not required for the *maintenance* of X inactivation. In human:rodent somatic cell hybrids in which only the human Xi chromosome is retained, silencing of X-linked genes can be maintained in the absence of *Xist* expression.

X inactivation in eutherian and marsupial mammals

The lineages leading to the eutherian and marsupial mammals diverged about 130 million years ago. Marsupials, like eutherians, use an XY (male):XX (female) sex determination system and a dosage compensation mechanism in which one of the two female X chromosomes is inactivated. However, the properties of the inactivated X are radically different in the two groups. They are compared in Table 12.1. The inactive X in marsupials is always the paternally derived homologue (Xp) whereas in eutherians the X to be inactivated in the embryo itself is chosen at random. X inactivation in eutherians occurs in all somatic tissues, whereas in marsupials there is variability between tissues and a sex chromatin body is seen only occasionally in marsupial cells but is common in eutherians.

Two of the most significant differences in terms of understanding possible molecular mechanisms of X inactivation are that, as yet, no marsupial homo-logue of *XIST* has been identified and the marsupial Xi shows no evidence of selective CpG methylation. The absence of an *XIST* homologue does not, of course, preclude the existence of a nonhomologous RNA that carries out the same function as an initiator of X inactivation. The divergence between euther-ian and marsupial *XIST* equivalents may be so great that significant homology is no longer detectable. Selective CpG methylation is thought to play a role in the stabilization of the inactive state of Xi and its absence from the marsupial Xi may contribute to its relative instability. In contrast to X inactivation in eutherian mammals, which is exceptionally difficult to reverse, reactivation of some genes on the marsupial Xi in tissue culture has been reported. There are only two properties that are unequivocally shared by the inactive X in eutherian and marsupial mammals. The first is that both replicate late in S-phase. The second is that both are marked by low levels of histone H4 acetylation.

In view of the male lethality of mutations that disrupt dosage compensation in other organisms, including other mammals, it is puzzling that incomplete dosage compensation is obviously tolerated in marsupials. The reason may lie in the fact that the marsupial X carries rather fewer genes than its eutherian coun-terpart. Only those genes on the long arm of the eutherian X are also present on the marsupial X, with those on the short arm being distributed on the auto-somes. A large part of the marsupial X is gene-poor and constitutively hete-rochromatic. Perhaps this reduction in gene number has made it possible for some cell types to tolerate a relaxation of dosage compensation, while not being sufficiently great to allow the organism to dispense with dosage compensation altogether.

Dosage compensation in *Drosophila*

The dosage compensation system in *Drosophila* exemplifies the second of the three methods of dosage compensation that were identified at the beginning of this chapter. This approach, namely doubling the rate of transcription of genes on the single male X, may seem almost the exact opposite to the one adopted by mammals, but the two systems have some problems in common. Both must dis-criminate between one chromosome (the X) and all the others (the autosomes), both are activated by a mechanism that depends on the X:autosome ratio and both must alter levels of gene expression across virtually an entire chromo-some. One obvious difference is that the dosage compensation mechanism in *Drosophila* is activated in males not females, and thereby avoids the tricky prob-lem of distinguishing between two homologous chromosomes in the same nucleus (i.e. of 'choosing' one X out of two, or more). Conversely, the fly is faced with the problem of exactly doubling the rate of transcription of genes on the male X, a quantitative change that could be seen as a more challenging puzzle than simply switching them all off.

Initiation of dosage compensation

The genetic switch that turns on the dosage compensation pathway in *Drosophila* is a gene called *sex lethal* (*Sxl*). The expression of *Sxl* and formation of a functional protein product (SXL) is regulated by a system that measures the ratio of X chromosomes to autosomes. Where there is one X chromosome per diploid set of autosomes (i.e. the X:A ratio is 0.5, as in males) functional SXL is not produced and where there are two (i.e. the X:A ratio is 1, as in females) it is. This is shown in Fig. 12.8. Some of the X-linked and autosomal genes involved (designated *numerators* and *denominators*, respectively) have been identified. With only one exception they encode transcription factors that control the activity of the *Sxl* promoter by interacting with upstream target sequences distributed over several kb.

The level of SXL protein produced is regulated by a complex system of alternative promoter usage and alternative splicing. Early in development (nuclear cycles 12–14) in embryos where the X:A ratio is 1.0 (i.e. females) there is a burst of *Sxl* transcription from the 'establishment' promoter, designated Pe (the target of the X:A counting system). Later in development *Sxl* transcription begins from a second 'maintenance' promoter, Pm. Where SXL protein is already present (i.e. in female cells), the transcript is spliced to give a functional SXL protein. Where SXL is absent (i.e. in male cells), the transcript is not correctly spliced and no SXL protein is produced. Thus, the system provides a feedback loop that maintains SXL in females but not in males (Fig. 12.9).

The absence of SXL initiates developmental pathways that lead to both male sexual characteristics and dosage compensation. In *Drosophila*, unlike mammals, the Y chromosome has no role in sex determination and XO flies are sterile, but otherwise normal, males. The SXL protein influences dosage compensation through its effect on *translation* of the transcript of a gene whose product is crucial for dosage compensation and male viability, namely *msl2* (see below). In the presence of SXL (i.e. in females), this transcript is not translated, with the result that no MSL2 protein is formed and dosage compensation does not occur. In mutant flies that are mosaic for *Sxl* expression, only those cells *without* SXL protein (assayed by immunofluorescence microscopy) contained male-specific components of the dosage compensation pathway.

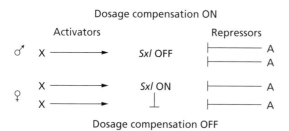

Fig. 12.8 The *sex lethal* (*Sxl*) gene in flies is regulated by the X:autosome ratio. Activators are products of genes on the X chromosome such as *sisterless* (*Sis*) A and B. These are also known as 'numerator' genes. Repressors are the products of autosomal genes such as *deadpan* (*Dpn*). These are also known as 'denominator' genes.

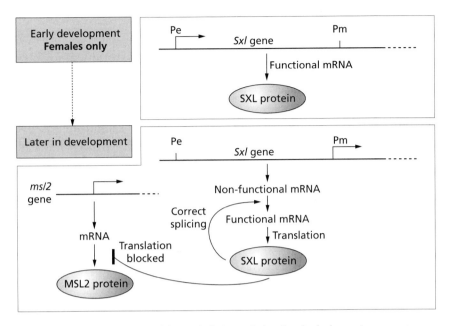

Fig. 12.9 Regulation of levels of the sex lethal protein involves both alternative promoters and alternative splicing.

Protein components of the *Drosophila* dosage compensation system

The list of components of the *Drosophila* dosage compensation system continues to grow and may not yet be complete (Table 12.2). The first components to be identified were the products of the male specific lethal genes *msl1*, *msl2*, *msl3* and *mle*. As the name suggests, male flies with loss of function mutations of these genes do not survive beyond the early stages of development whereas females are unaffected. All four of these proteins have been shown, by immunofluorescence microscopy, to associate with several hundred specific sites on the X chromosome in male, but not female, cells (Plate 5.1). It seems, therefore, that the MSL proteins have a direct effect on the structure and function of the male X chromosome. The fact that all four are located at the same X chromosome sites suggests that they act in a co-ordinated manner, possibly as a multi-subunit complex. This possibility is supported by the finding that different MSL proteins co-immunoprecipitate, i.e. an antibody to one protein will precipitate all of them at the same time. Surprisingly, of the four MSLs, three are present in female cells, either at similar levels to males (MLE) or at 5–10% male levels (MSL1 and MSL3). Furthermore, there is evidence to suggest that reduced levels of MSL1 and MSL3 in females may result from their more rapid degradation, possibly as a result of their failure to form a stabilizing multiprotein complex. Only MSL2 is truly male-specific. The most recently identified protein component of the dosage compensation complex is the histone acetyltransferase, MOF (*Males absent On the First*). The enzyme is encoded by a gene on the

Table 12.2 Components of the dosage compensation complex in *Drosophila melanogaster*.

Component (gene)	Gene location	Present in		On male X	Putative function
		males	females		
MLE (*maleless*)	2	+	+	+	RNA helicase, chromatin remodelling
MSL1 (*male specific lethal 1*)	2	+	+/−	+	Recognition of chromatin entry sites (as complex with MSL2)
MSL2 (*male specific lethal 2*)	2	+	−	+	Stabilization of MSL1 through RING finger; recognition of chromatin entry sites
MSL3 (*male specific lethal 3*)	3	+	+/−	+	RNA recognition by chromodomain?
MOF (*males absent on the first*)	X	+	+	+	Histone acetyltransferase, specific for H4 lysine 16
roX1 (*RNA on the X 1*)	X	+	−	+	Locates to *roX2*-defined 'entry points', aids DCC assembly
roX2 (*RNA on the X2*)	X	+	−	+	Targeting of 'entry points' on the male X, assembly of DCC

X (first) chromosome, making the detection of mutants more difficult than for the other (autosomal) components. MOF may not be the only component of the complex with chromatin remodelling activity; MLE has a domain homologous to the ATPases found in helicases and has both ATPase and RNA/DNA helicase activities *in vitro*.

Noncoding RNAs are essential for dosage compensation in *Drosophila*

In addition to these proteins, the complex contains two noncoding RNAs encoded by genes on the X chromosome, *roX1* and *roX2*. The RNA transcripts are found only in males. Deletion of *roX1* alone has no effect on the phenotype of male or female flies. It is probable that loss of both *roX1* and *roX2* abolishes dosage compensation and is lethal in males. (We cannot be sure because there is not yet a simple mutation that eliminates *roX2*. The only *roX2*- lines available have a deletion that also removes adjacent genes that are essential for normal development.) Intriguingly, these RNAs seem to play a role in the targeted distribution of the dosage compensation complex along the X chromosome. If the *roX1* or *roX2* genes are inserted into an autosome, then MSL proteins assemble at the insertion site and the complex can spread into the surrounding chromatin (Fig. 12.10). The degree of spreading is substantially greater at some insertion sites than others, but also varies from one nucleus to another in the same fly (a phenomenon reminiscent of position effect variegation, PEV, discussed earlier).

It has been possible to use flies deficient in one or another components of the

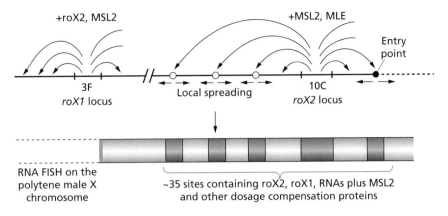

Fig. 12.10 Spreading of the *roX* non-coding RNAs across the *Drosophila* male X chromosome involves specific entry points. Entry points may be marked by components of the dosage compensation complex. Experiments have shown that MSL2 and other components of the complex are necessary for *roX2* RNA to reach the entry points, from where it can spread locally. (See Meller *et al.* 1997, 2000.)

complex to study targeting of *roX* RNAs to the X chromosome. In a series of complex and elegant experiments, it has been shown that there are between 30 and 40 'entry sites' on the polytene X chromosome in third instar larvae, to which *roX* RNAs are targeted. The *roX2* RNA seems to play the leading role in the targeting process. This is summarized in Fig. 12.10. In the *absence* of MLE, *roX2* RNA is still detected at its site of synthesis (band 10C), but not anywhere else. In the *presence* of MSL proteins (MSL3 seems not to be necessary), *roX2* RNA is detected at each of 30–40 sites on the X chromosome, the 'entry sites'. The *roX1* transcript is not needed for this, but *roX1* RNA is assembled, along with MSL proteins, at the *roX2*-containing entry sites and the complex then spreads locally from each site.

The *roX* RNAs illustrate two radically different types of spreading. The first, in which the RNA spreads continuously in *cis*, from its site of synthesis, is reminiscent of the sort of spreading discussed before in the context of position effect variegation and silencing. The second is new and potentially much more flexible. It allows spreading over large distances that can miss out intervening regions of the genome. Spreading can even occur in *trans*, in that *roX2* expressed from a transgene inserted on an autosome can be targeted to entry sites on the X, from which it then spreads in *cis*. Whether the signal that defines an entry site resides in its DNA sequence or in an epigenetic modification remains to be determined. Nonetheless, the entry site hypothesis explains earlier observations about the persistence or otherwise of dosage compensation on translocated chromosomes. For example, fragments of the X chromosome translocated to autosomes are often, but not always, dosage compensated. Whether they are or not may depend on whether or not they carry *roX* entry sites. Similarly, whether fragments of autosome translocated to the X are dosage compensated or not will depend on their proximity to an entry site and to their ability to resist the local spreading of the complex. (Recall the variability of spreading from autosomal *roX* transgenes depending on their chromosomal location.)

Differences in chromatin structure between male and female X chromosomes

In *Drosophila*, there is direct microscopical evidence that the chromatin structure of male and female X chromosomes is different. In polytene chromosome squashes from third instar female larvae, the two X chromosome homologues are aligned side-by-side to form what appears to be a single polytene chromosome. This should, in theory, be twice as thick as the male polytene X. In fact the polytene X chromosomes in male and female cells are of similar thickness, the reason being that the male X has a visibly more diffuse structure. The two chromosomes also differ in a more specific feature of their chromatin, namely the pattern of acetylation of histone H4. By immunostaining of polytene chromosome squashes with antisera to acetylated H4, it was shown that the male X chromosome was marked uniquely by a high level of a specific acetylated isoform, namely H4 acetylated at lysine residue 16, H4Ac16. Significantly, this H4 isoform was distributed along the male X in a pattern of discrete bands that corresponded almost exactly to the bands containing MLE and other protein components of the dosage compensation pathway (Plate 5.1). It is important to note that the male X chromosome is not *generally* hyperacetylated. It is indistinguishable from autosomes in the intensity with which it labels with antibodies to H4 acetylated at lysines 5, 8 and 12.

Histone acetylation and dosage compensation

H4Ac16 is a highly conserved component of the *Drosophila* dosage compensation system, being found not only in *D. melanogaster*, but also in distantly related species such as *D. pseudoobscura* and *D. miranda*. However, there is as yet little evidence as to how H4Ac16 might be incorporated into the dosage compensation complex on the male X. Analysis of loss-of-function mutants of each of the four MSLs has shown that H4Ac16 is seen on the male X only when the dosage compensation system is complete and functional. H4Ac16 cannot therefore be a primary and independent signal that serves to initiate assembly of the complex.

Unfortunately, until further experimental evidence accumulates, there remain several equally likely roles for H4Ac16 on the male X. It may serve to: (i) open up the higher-order chromatin structure of the male X (either directly or indirectly) and thereby facilitate binding of transcription factors and other proteins; (ii) alter DNA–H4 contacts as part of the (unknown) mechanism that brings about the doubling of transcriptional activity of X-linked genes in male cells; (iii) tether the dosage compensation complex to chromatin; (iv) provide an interaction site for nonhistone proteins involved in transcriptional regulation; or (v) provide a marker that maintains the location of the dosage compensation complex from one cell generation to the next. These possibilities are not mutually exclusive.

Less mysterious is the question of how H4Ac16 comes to be located specifically on the male X chromosome. Recent experiments have shown the histone acetyltransferase MOF, an integral part of the dosage compensation complex, is exquisitely specific for H4 lysine 16. MOF is present in both male and female cells (Table 12.2), but is targeted to the X chromosome only in males, presumably by components of the dosage compensation complex.

How is transcription increased just twofold?

In most of the gene regulation problems considered so far in this book, the challenge has been to explain how specific genes are switched on or off. In *Drosophila* dosage compensation the problem is quite different, namely how can we increase the transcription of many different genes (namely all X-linked genes in males) exactly twofold? Enormous progress has been made in identifying the components of the dosage compensation complex, how they are assembled and how their levels are regulated in males and females. But the question of how the complex adjusts the level of transcription remains a mystery. Two components are potentially important in this respect. One is MLE, which possesses an ATPase domain of the type found in helicases and components of chromatin remodelling complexes (Chapter 8). The other is MOF, a histone acetyltransferase. HATs are frequently found as part of transcription initiation complexes and, like helicases, have the potential to alter chromatin structure.

So, are these activities involved in remodelling chromatin at X-linked promoters and thereby altering the efficiency with which transcription initiates? Probably not, for the simple reason that it is difficult to envisage how the various different types of promoter along the X chromosome could all be influenced in just the same way by the dosage compensation complex in male cells. More likely is the possibility that the complex influences the rate of transcription post-initiation, i.e. the rate at which the core polymerase progresses along the gene. Chromatin modifying activities could still be involved, for example by facilitating displacement of nucleosomes ahead of the polymerase. The *roX* RNAs might also play a role, possibly by interacting with the new transcripts. Both of these possibilities are attractive in that they operate at a stage in the transcription process that is common to all genes. Individual genes would still show their own patterns of regulation, governed by rates of initiation at the promoter, but these would be overlain by downstream effects on polymerase progression.

Dosage compensation in *C. elegans*

The nematode worm *C. elegans* has become a popular model organism for studying the molecular biology of many fundamental cellular processes. The justification given for spending so much time and money on the study of worms is usually that the processes under examination are highly conserved through evolution and the results obtained in this simple animal are directly applicable to more complex organisms, such as ourselves. This line of reasoning has been amply justified by the exciting discoveries that have originated from work with *C. elegans*, particularly in the fields of developmental biology and apoptosis (programmed cell death). Studies of dosage compensation in worms have also proved enlightening, although the insights into the process as a whole have come as much from the contrasts between worms and other organisms as the similarities.

Dosage compensation mechanisms have evolved rapidly and, as noted at the beginning of this chapter, mammals, flies and worms all use fundamentally different mechanisms to achieve the same end. In worms, even the sexes are different. The two sexes are distinguished by a chromosomal difference, but in worms the sex with just one X chromosome is male (there is no Y), while that with two X chromosomes is hermaphrodite, i.e. has both male and female germ cells. For studies of dosage compensation, this is not particularly important in that the hermaphrodites are essentially females that can make a limited number of sperm for self fertilization if no males are available. (This characteristic is useful for genetic experiments in that worms can be self-fertilized, like plants.) As with flies, sex is determined by the X:autosome (X:A) ratio. Animals with a ratio of 0.5 are males and those with a ratio of 1.0 are females/hermaphrodites. In what follows, purely in the interests of simplicity, XX worms will be referred to as females.

In worms, dosage compensation is achieved by reducing the transcription of genes on each of the two X chromosomes in females. This approach resembles that used by mammals in the sense that it operates through females. It is simpler than the mammalian mechanism in that it does not need to distinguish between two essentially identical chromosomes in the same cell (inactivating one but not the other). It does, however, face the additional problem of requiring a reduction in transcription of close to 50%, rather than a complete shut-down.

Switch genes and proteins in *C. elegans*

As in *Drosophila*, there is a crucial switch gene in *C. elegans* that responds to the X:A ratio. This gene, *xol-1*, was initially identified, as were the other crucial genes described below, through genetic studies of mutants showing defects in sex determination or dosage compensation or both. *xol-1* mutants are lethal in males, but without effect in females. The XOL-1 protein is present at high levels in males and low levels in females (Fig. 12.11). In worms, a few of the X-linked and autosomal genes (the numerators and denominators) that regulate the *xol-1* switch (Fig. 12.11) have been identified genetically and two of the X-linked numerator (repressor) elements have been further characterized. The *sex-1* protein product (i.e. SEX-1) represses *xol-1* transcription by binding upstream of the transcription start site, while the FOX-1 protein acts post-transcriptionally (either on RNA processing or translation), thereby reducing production of XOL-1 protein still further. Because SEX-1 and FOX-1 act at different points in the pathway from gene to protein, their effects are synergistic, something that helps make the level of XOL-1 so exquisitely sensitive to the X:A ratio. The *xol-1* switch still works when chromosome numbers are manipulated to give X:A ratios of 2X:3A (males) and 3X:2A (females).

XOL-1 operates by repressing the transcription of a second gene, *sdc-2*. The protein product of this gene is involved directly in regulating transcription of X-linked genes and is also necessary for regulating levels of the protein product of a second essential gene, *sdc-3*. (Exactly how the levels of the SDC-3 protein are regulated by SDC-2 remains uncertain.) Thus, in XO males in which XOL-1 levels are high, *sdc-2* is off, SDC-3 levels are low and dosage compensation does not occur. Conversely, in XX females, XOL-1 levels are low, *sdc-2* is on, SDC-3 levels are high and dosage compensation can operate (Fig. 12.11). Interestingly, in light of the behaviour of the *Sxl* gene in flies, a burst of *xol-1* transcription very early in development (from the 28-cell stage through to gastrulation) is enough to trigger irreversibly both sex determination and dosage compensation. After this stage XOL-1 is no longer necessary (unlike *Sxl* in flies).

The down-regulation of transcription of X-linked genes in female *C. elegans* is achieved by the action of a multiprotein complex associated with the two female X chromosomes. The components of this complex are shown diagrammatically in Fig. 12.12. They have been identified by a combination of approaches. Genetic analysis of spontaneous and induced mutants has identified crucial components. Immunoprecipitation has shown that antibodies to one

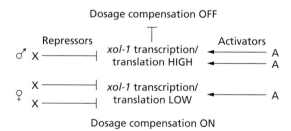

Fig. 12.11 The *xol-1* gene in worms is regulated by the X:autosome ratio. The XOL-1 protein product represses genes such as *sdc-2* and *sdc-3* that are necessary for dosage compensation. Known X-linked repressors (numerators) include the genes *sex-1* and *fox-1*, whose protein products repress, respectively, either the transcription of *xol-1* or its processing and translation. Autosomal activators (denominators) have yet to be identified.

component of the complex can precipitate other components as well and im-muno-microscopy has shown that all components of the complex are located on the X chromosomes of female, but not male, animals. As in *Drosophila*, several of the components of the complex are found in both sexes, but only in females do they associate with the X chromosomes. Association with the interphase X chromosomes must require one or both of the two female-specific components, SDC-2 and SDC-3.

A link between dosage compensation and chromosome condensation

Circumstantial evidence suggests that the dosage compensation complex achieves the required 50% down-regulation of transcription by a chromatin condensation mechanism. This tentative conclusion rests on the functional properties of three components of the complex, namely DPY-26, DPY-27 and MIX-1. The strongest evidence comes from studies of MIX-1. This protein (along with DPY-27) is a member of a highly conserved family, the SMC (structural maintenance of chromosomes) proteins. These proteins are necessary for the condensation of chromatin, and concomitant transcriptional silencing, that occurs as cells enter mitosis (see Chapter 5). Furthermore, MIX-1 seems to carry out just this function. In both male and female worms, MIX-1 has been shown, by immunomicroscopy, to associate with condensed mitotic chromosomes. Once mitosis is complete, the protein dissociates from the chromosomes. In males this dissociation is accompanied by degradation, but in females MIX-1 is stabilized by components of the dosage compensation complex (specifically DPY-26 and DPY-28) and targeted to the X chromosomes. Crucially, loss of function mutations of *mix-1* disrupt mitosis, providing direct evidence that MIX-1 plays a role in this process.

MIX-1 may not be the only component of the dosage compensation complex that has a second function. DPY-26 is associated with meiotic chromosomes in both male and female germ cells and *dpy-26* mutations show impaired meiotic function. Thus, both MIX-1 and DPY-26 are proteins with crucial functions in fundamental cellular processes that have been conscripted into the dosage compensation mechanism. The use of such proteins may explain how

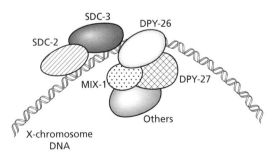

Fig. 12.12 Components of the *C. elegans* dosage compensation complex.

dosage compensation has been able to evolve so rapidly and it will be surprising if the mechanisms in flies and mammals have not made use of the same strategy.

Lessons from dosage compensation

Although the three dosage compensation mechanisms described in this chapter are, at first sight, very different, they share some revealing properties. Mechanisms of dosage compensation seem to have evolved rapidly, perhaps driven by the progressive degeneration of one chromosome in the heterogametic sex (Fig. 12.1). Therefore, mechanisms have been assembled using pre-existing components. This is most apparent in *C. elegans*, where SMC proteins involved in chromosome condensation at mitosis have been pressed into action. But it is likely also to be the case for the protein components of the *Drosophila* complex, all of which, with the exception of MSL2, are expressed in both males and females. Intriguingly, the most likely components to have a specialized role exclusively in dosage compensation are the noncoding RNAs *Xist* and *Tsix* in mammals and *roX1* and *roX2* in flies. All of them are expressed from the X chromosome and in only the sex in which it is dosage compensated. *Xist, roX1* and *roX2* are all involved in the initiation and spreading of the dosage compensation signal. Perhaps, in evolutionary terms, it is quicker to put in place a new, functional RNA than a new protein.

Perhaps the most important new concept to come out of this journey through the world of dosage compensation is the idea that noncoding RNAs can mediate the spread of a transcriptional signal both in *cis* and in *trans*. Both *roX* and *Xist* RNAs can spread locally from their sites of synthesis, either from the endogenous X-linked gene or an autosomal transgene, and *roX2* can be transported to new sites on the X by a mechanism that requires MSL proteins. Can *Xist* be transported in the same way? It would certainly explain its ability to avoid certain regions of the X chromosome, such as the PAR or certain types of centric heterochromatin. It might also explain why some genes are not silenced on the inactive X. Microscopical analysis of metaphase chromosomes lacks the detailed resolution possible with interphase polytene chromosomes, so we cannot tell whether the distribution of *Xist* RNA resembles the banded distribution of *roX1* and *roX2*. However, the ability of noncoding RNAs to transfer a transcriptional signal over large genomic distances is of such fundamental importance that it will be surprising if answers to these questions are not found rapidly.

Dosage compensation also illustrates the importance of distinguishing between mechanisms for *establishing* and *maintaining* a particular state of transcriptional activity (discussed in Chapter 10). Dosage compensation is essential for viability and must last for the life of the organism, so both mechanisms are crucial. X inactivation does not occur in cells lacking a functional *Xist* gene.

However, if the *Xist* gene is removed from (hybrid) cells in which an inactive X chromosome is already established, then the inactive state is maintained. So *Xist* is necessary for establishment, but not maintenance. In contrast, the overall deacetylation of the inactive X, and methylation of CpG islands, occurs, in development, only *after* silencing of X-linked genes and the increase in *Xist* RNA. These changes are likely to be part of the maintenance mechanism. It is possible that H4Ac16 on the dosage compensated male X in *Drosophila* is also part of the mechanism by which a particular chromatin structure is maintained from one cell generation to the next. Could it act, like H3 methylated at lysine 9 in heterochromatin (Chapter 11), as a regional binding site for non-histone proteins that modulate or maintain chromatin structure? And what role does histone methylation play in the formation and maintenance of the facultative heterochromatin of the inactive X? Many questions remain, but it is becoming increasingly clear that maintenance of dosage compensation involves several different components, acting in a complementary or synergistic fashion, to lock in place a state of transcriptional activity so firmly that its reversal is all but impossible.

Further reading

Reviews

Baker, B. S., Gorman, M. & Marin, I. (1994) Dosage compensation in *Drosophila. Ann. Rev. Genet.*, **28**: 491–521.

Charlesworth, B. (1996) The evolution of chromosomal sex determination and dosage compensation. *Curr. Biol.*, **6**: 149–162.

Cline, T. W. & Meyer, B. J. (1996) Vive la différence: males vs. females in flies vs. worms. *Ann. Rev. Genet.*, **30**: 637–702.

Graves, J. A. M. (1994) The origin and function of the mammalian Y chromosome and Y-borne genes—an evolving understanding. *BioEssays*, **17**: 311–320.

Heard, E., Clerc, P. & Avner, P. (1997) X-chromosome inactivation in mammals. *Ann. Rev. Genet.* **31**: 571–610.

Meyer, B. J. (2000) Sex in the worm: counting and compensating X-chromosome dose. *Trends Genet.*, **16**: 247–253.

Riggs, A. D. & Pfeifer, G. P. (1992) X-chromosome inactivation and cell memory. *Trends Genet.*, **8**: 169–174.

Strachan, T. & Read, A. P. (1966) *Human Molecular Genetics*. Bios Scientific Publishers, Oxford.

Selected papers

Dosage compensation in mammals, the X inactivation centre, Xist and Tsix

Clemson, C. M., McNeil, J. A., Willard, H. F. & Lawrence, J. B. (1996) *Xist* RNA paints the inactive X chromosome at interphase: Evidence for a novel RNA involved in nuclear/chromosome structure. *J. Cell Biol.*, **132**: 259–275.

Duthie, S. M., Nesterova, T. A., Formstone, E. J. *et al.* (1999) *Xist* RNA exhibits a banded localization on the inactive X chromosome and is excluded from autosomal material *in cis*. *Hum. Mol. Genet.*, **8:** 195–204.

Eggan, K., Akutsu, H., Hochedlinger, K. *et al.* (2000) X-chromosome inactivation in cloned mouse embryos. *Science*, **290:** 1578–1581.

Herzing, L. K. B., Romer, J. T., Horn, J. M. & Ashworth, A. (1997) *Xist* has properties of the X chromosome inactivation centre. *Nature*, **386:** 272–289.

Lee, J. T. (2000) Disruption of imprinted X inactivation by parent-of-origin effects at *Tsix*. *Cell*, **103:** 17–27.

Lyon, M. F. (1961) Gene action in the X chromosome of the mouse (*Mus musculus* L). *Nature*, **190:** 373.

Marahrens, Y., Panning, B., Dausman, J., Strauss, W. & Jaenisch, R. (1997) *Xist* deficient mice are defective in dosage compensation but not spermatogenesis. *Genes Dev.*, **11:** 156–166.

Rastan, S. & Brown, S. D. M. (1990) The search for the mouse X chromosome inactivation centre. *Genet Res.*, **56:** 99–106.

Chromatin of the inactive X

Constanzi, C. & Pehrson, J. R. (1998) Histone macroH2A1 is concentrated in the inactive X chromosome of female mammals. *Nature*, **393:** 599–601.

Eils, R., Dietzel, S., Bertin, E., Schrock, E. & Speicher, M. R. (1996) Three-dimensional reconstruction of painted human interphase chromosomes: active and inactive X chromosome territories have similar volumes but differ in shape and surface structure. *J. Cell Biol.*, **135:** 1427–1440.

Hansen, R. S., Canfield, T. K. & Gartler, S. M. (1996) Role of late replication timing in the silencing of X-linked genes. *Hum. Mol. Genet.*, **5:** 1345–1353.

Jeppesen, P. & Turner, B.M. (1993) The inactive-X chromosome in female mammals is distinguished by a lack of H4 acetylation, a cytogenetic marker for gene expression. *Cell*, **74:** 281–289.

Keohane, A. M., O'Neill, L. P., Belyaev, N. D., Lavender, J. S. & Turner, B. M. (1996) X-inactivation and H4 acetylation in embryonic stem cells. *Dev. Biol.*, **180:** 613–630.

Dosage compensation and the male X chromosome in Drosophila

Akhtar, A. & Becker, P. B. (2000) Activation of transcription through histone H4 acetylation by MOF, an acetyltransferase essential for dosage compensation in *Drosophila*. *Mol. Cell*, **5:** 367–375.

Bone, J. R., Lavender, J. S., Richman, R. *et al.* (1994) Acetylated histone H4 on the male X chromosome is associated with dosage compensation in *Drosophila*. *Genes Dev.*, **8:** 96–104.

Dobzhansky, T. (1957) The X-chromosome in the larval salivary glands of hybrids *Drosophila insularis* × *Drosophila tropicalis*. *Chromosoma*, **8:** 691–698.

Meller, V. H., Gordadze, P. R., Park, Y. *et al.* (2000) Ordered assembly of *roX* RNAs into MSL complexes on the dosage-compensated X chromosome in *Drosophila*. *Curr. Biol.*, **10:** 136–143.

Meller, V. H., Wu, K. H., Roman, G., Kuroda, M. I. & Davis, R. L. (1997) roX1 RNA paints the X chromosome of male *Drosophila* and is regulated by the dosage compensation system. *Cell*, **88:** 445–457.

Turner, B. M., Birley, A. J. & Lavender, J. (1992) Histone H4 isoforms acetylated at specific lysine residues define individual chromosomes and chromatin domains in *Drosophila* polytene nuclei. *Cell*, **69:** 375–384.

Dosage compensation in C. elegans

Carmi, I., Kopczynski, J. B. & Meyer, B. J. (1998) The nuclear hormone receptor SEX-1 is an X chromosome signal that determines nematode sex. *Nature*, **396:** 168–173.

Lieb, J. D., Albrecht, M. R., Chuang, P. T. & Meyer, B. J. (1998) MIX-1: an essential component of the *C. elegans* mitotic machinery executes X chromosome dosage compensation. *Cell*, **92:** 265–277.

Index

Note: page numbers in *italics* refer to figures; those in **bold** refer to tables.